NEWTON
PUBLIC
LIBRARY
Gift
of
Florence Bessmer
Foundation

CLONING and the NEW GENETICS

CLONING and the NEW GENETICS

Margaret O. Hyde &
Lawrence E. Hyde

ENSLOW PUBLISHERS, INC.

Bloy St. & Ramsey Ave.
Box 777
Hillside, N.J. 07205
U.S.A.

P.O.Box 38
Aldershot
Hants GU12 6BP
U.K.

Library of Congress Cataloging in Publication Data:

Hyde, Margaret Oldroyd, 1917-
Cloning and the new genetics.

Bibliography: p.
Includes index.
1. Cloning. 2. Genetics. I. Hyde, Lawrence E. II. Title.
QH442.2.H93 1984 575.1 83-20727
ISBN 0-89490-084-6

Printed in the United States of America

10 9 8 7 6 5 4

For Molly Reeder Hyde

Contents

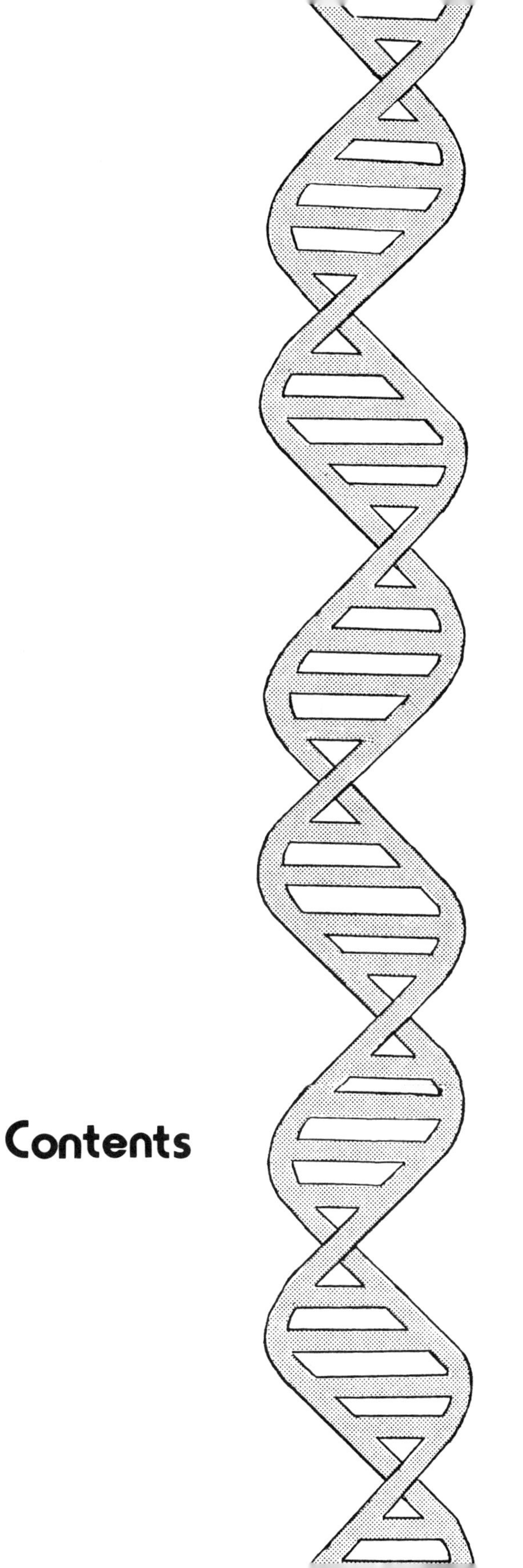

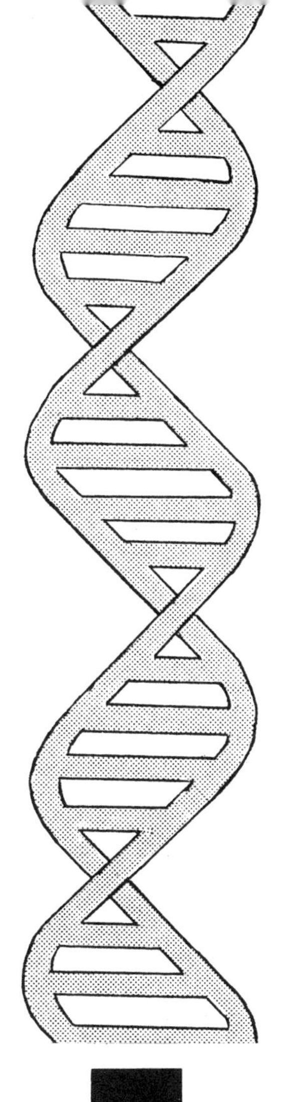

1 What Is a Clone?

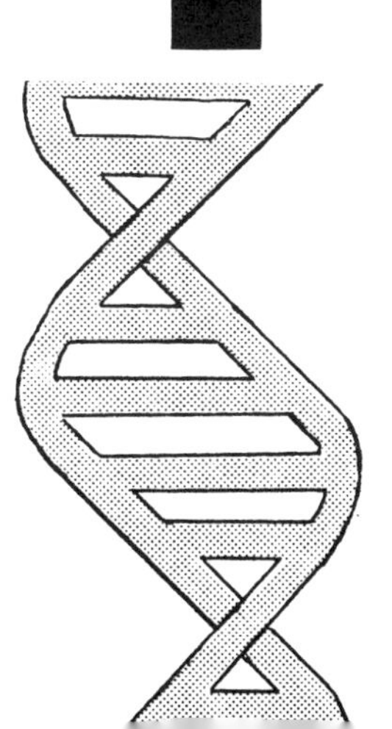

How would you feel if someone called you a clone? Is this a compliment or an insult? Suppose some friends said that they would like to clone you. Would you think this idea was a good one?

A clone is a group of identical cells or organisms that are descended from a common ancestor through asexual reproduction. In asexual reproduction, the offspring come from only one parent, so all cells in the clone contain the same genetic material. The members of a clone may be regarded as extensions of a single individual.

The fact that every living thing is made of one or more building blocks known as cells has been known for a long time. Usually, a cell is confined by a cell wall and contains a nucleus in which there are genes. Cytoplasm, a jellylike substance, surrounds the nucleus. The human body consists of about 60 trillion of these living units, and all but certain reproductive cells contain a full complement of genes. Estimates range from 30,000 to 100,000 genes per cell. The enormous complexity and diversity of the human body and all living things is based on information stored in genes and later expressed.

Cloning is not new, but many people think it is something that scientists have just discovered and are about to use in the wrong way.

When authors of science fiction want to clone a male, they use a cell from the body of a man or boy; when authors want to clone a female, they use a body cell from a girl or woman. In science fiction, one can just take a cell from the person one wants to copy.

Many authors of science fiction make use of human clones to entertain through fantasy and fear. In the film *Sleeper*, Woody Allen used a plot that revolved around "cloning the nose." *The Boys From Brazil* introduced unpleasant characters as clones of Hitler. In the television show *Computercide*, a cloned arm that was part of a museum exhibit could be activated by rubbing it. The hand which was fastened to this detached arm closed to crunch a nut in its palm. Upstairs in this so-called science building, cloned people were maintained in cases by a system of tubes and liquids that circulated around them. All of this indicated that the cloning of men had not been perfected, but this science fiction took place in the 1990s, and one might expect "better" results in the future. Cloning of people is easy in fiction, but it is far different from natural cloning and the kind of cloning that is being used today by genetic engineers.

Cloning is an important part of the biological revolution; its future impact has been compared with the introduction of the silicon chip to electronics. It has been the subject of much controversy, partly because the public image of cloning has been distorted by science fiction. But scientific cloning has great potential value in studies relating to cancer, aging, genetic diseases, and agriculture.

In one kind of laboratory cloning, scientists are duplicating genes. They remove genes from a chromosome of one kind of plant or animal and insert them into a chromosome of another species through the exciting and complicated process of gene splicing, which is described in Chapter 4.

In another kind of laboratory cloning, scientists are combining, or fusing, two kinds of cells to make a desired third kind, called a hybridoma; then the new cell is cloned. This technique, which plays a large role in cancer research, is described in Chapter 6.

The term "clone" has long been used in horticulture to designate all the descendents of a single plant produced by vegetative methods such as grafting, leaf root and stem cuttings, and layering. So when you eat a MacIntosh or a Delicious apple, you are eating a clone. Peaches, strawberries, carnations, potatoes, roses, grapes, and many other plants are clones when they are reproduced by vegetative methods. Some of the European varieties of grapes are clones of plants that originated as much as 2000 years ago. The desirable qualities of certain wine grapes have been preserved by growing new plants from the old ones through cloning, which gives them the same heritage.

A common example of cloning is the propagation of the common orange day lily. The orange day lily has been widely reproduced in gardens by dividing the basal parts of old plants and planting the sections. This method is commonly used since

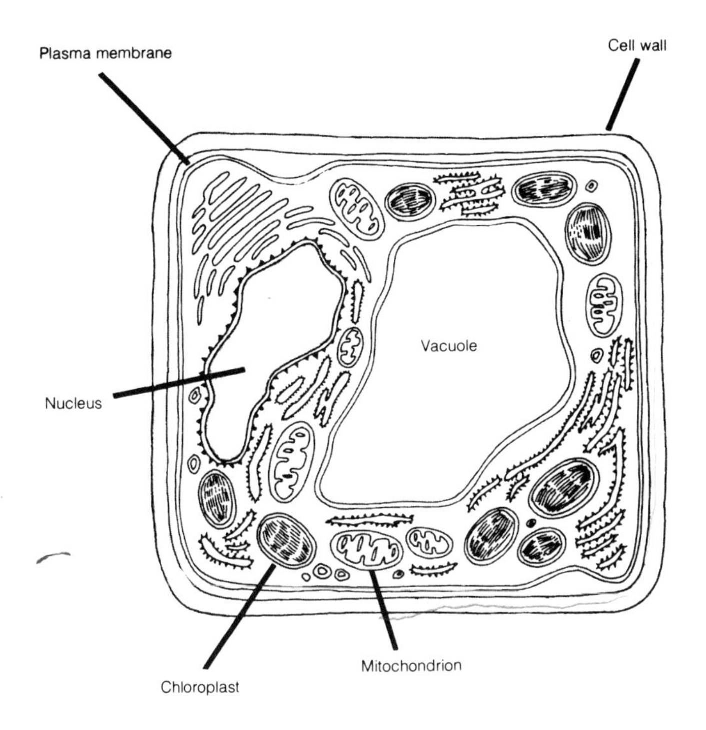

Figure 1-1. A typical plant cell, one thousand times larger than a bacterial cell.

Figure 1-2. Strawberry plants that have reproduced by cloning.
USDA Agricultural Research Service

none of the plants produces seed when pollinated with its own pollen. Hybrid seed can be produced by pollinating one day lily using pollen from a different day lily plant. But it is probable that most of the cultivated lilies and those that grow wild in the fields are clones of one individual plant that lived about four hundred years ago.

One of the fears about cloning is the possibility that an unlimited number of undesirable organisms may grow out of control. There are cases of natural cloning in which this has happened. Water hyacinths produce seeds, but they spread mainly through natural vegetative propagation of underwater stems. In one eight-month growing season, ten of these plants could produce more than half a million plants, enough to cover acres of water surface and block navigation.

A variety of blueberries that grows wild has been reported to cover an area half a mile in diameter. Since these blueberries reproduce by sending out underground stems at the rate of one foot per year, the clones from the plants in this area are believed to be about a thousand years old.

Cattails, milkweed, bindweed, and many other weeds produce clones at a rapid rate. Some woody plants too, such as Osage orange, sumac, lilac, and beech, are often formed by aerial stems that arise from underground roots. So cloning often produces thick groves of woody plants.

Laboratory cloning of plants is not as old as natural cloning, and it did not cause much excitement when it began. Professor F.C. Steward was director of the Laboratory for Cell Physiology, Growth and Development at Cornell University in 1958 when he took cells from a carrot root, placed them in a rotating tube where they were bathed in nutrient solution, and found that tissue began to grow. In less than three weeks, the carrot cells had multiplied about eightyfold. Steward continued with various experiments in the cloning of carrot cells, and he found that some of the masses broke away from the main mass. Some of these formed clumps that produced roots, and when transferred to a solid medium, they put up shoots. This discovery is now famous in the history of engineered reproduction.

The cloning of plants is far more than a laboratory curiosity today. The techniques developed by researchers in the 1950s have been expanded so that most orchids, Boston ferns, oil palms, African violets, and a large number of forest trees are now cloned. They are grown by isolating single cells, then putting the cells in an appropriate medium where they divide and form the numerous cells that make up young plants. This method, known as tissue culture, is a valuable form of laboratory cloning. When the single cells are given the proper hormones and correct growing conditions, plants grow through intermediate steps and finally become mature plants. This kind of cloning combined with genetic engineering is described in a later chapter.

Cloning of animals is a subject that causes far more controversy than the cloning of plants, although there is some natural cloning of animals. Scientists once believed that vertebrate animals, those with backbones, did not have the capacity to clone. However, in 1934, Carl and Laura Hubbs, a team working at the University of Michigan, published a report on an all-female fish population that they discovered in the waters of northern Mexico. This species, known commonly as the Amazon molly, is related to the popular tropical mollies found in the aquariums of many hobbyists. Their study provided the first hard evidence of clonal reproduction in fish. Since that time, many species of fish and amphibians that reproduce asexually have been discovered.

In fish and amphibians, there are three methods by which natural clones are produced. In two cases, the males of the species participate in the breeding process. In one method, the male's genes have an effect in only the first generation. For example, the common green frog of Central Europe is a cross between a species that lives in the water and one that lives on land. However, the genes from the land ancestor do not survive in the offspring. In the second method, the sperm serves only to stimulate the cell division and the development of the egg. There is no merger of chromosomes, the carriers of the genes, from the father and the mother; they are all maternal. In parthenogenesis, the third method of natural cloning, no males are involved. The eggs mature without any contribution from the male.

Numerous studies have been made in effort to determine whether or not some lizards reproduce by parthenogenesis, or whether there might be some sperm involved. For example, female lizards can store sperm for months, and they may have been in contact with males before they were captured. At least one hundred scientific papers have appeared on this subject, but questions remain. Of the estimated 3000 species of lizards in the world, only about 30 different species appear to reproduce asexually. Snakes are close relatives of lizards, and it is possible that one variety of snake, *Typhlina bramina*, which measures only six inches when fully grown, may reproduce clonally.

The nine-banded armadillo, a mammal, reproduces asexually routinely, producing offspring that are identical quadruplets. Another variety of armadillo, the mulita armadillo, produces from eight to twelve identical young.

Human offspring are often called clones when they are identical twins or triplets because they are multiple births from the same fertilized egg. Some scientists object to the use of the word clone for such offspring, because they are clones of each other, not of their parents, having begun from the one fertilized egg.

Is the laboratory cloning of animals possible? Dr. Robert Briggs and Dr. Thomas King performed such an experiment on frogs as long ago as the early 1950s. They removed nuclei from cells of frog embryos in various stages of development. Then they put each nucleus into an activated frog egg whose own nucleus had been removed. Eggs with nuclei from frog embryos in early stages of development grew into normal frogs. However, eggs with nuclei from embryos at more advanced stages either grew abnormally or stopped growing. Repeated nuclear transplant experiments indicated that nuclei from embryos in early stages contained the full instructions for creation of a whole frog; in later stages certain genes were turned on or off to cause development of specific parts of a frog.

Dr. J. B. Gurdon and his associates at Oxford University cloned African clawed frogs from well-differentiated parts of the embryo. They were able to take the nucleus from an intestinal cell of a fully developed tadpole and transplant it into an egg cell whose nucleus had been removed. This egg cell produced a normal frog. But in these experiments, too, there were limits in the age of the cell that could be used successfully in the cloning.

The renucleated egg carries all of the genetic information from the animal from which the nucleus was taken. The animal that grows from this egg has the same genetic features as the donor, so one could theoretically make as many copies as were desired.

At the University of Oregon, George Streisinger and his colleagues have cloned zebra fish, the black striped freshwater fish found in many aquariums. In their efforts to learn more about hereditary problems that affect the nervous system development in animals that have a backbone, these molecular biologists chose the zebra fish as an experimental animal because these small fish develop rapidly, producing several hundred eggs every three weeks.

The cloning process begins by removing eggs from the female fish and fertilizing them with sperm that have been irradiated with ultraviolet light. The fertilized eggs are then placed in a bottle and either heated to 106 degrees Celsius or subjected to underwater pressure of 8000 pounds per square inch. This action prevents the first cell division. Although the male's contribution to hereditary characteristics has been eliminated by the exposure to ultraviolet light, the sperm do activate the eggs. The egg cells produce two sets of identical chromosomes because of the heat or pressure which prevents the first cell division. Only about 20 percent of the eggs that are treated survive, but the scientists have been successful in producing clones that consist of 200 identical fish.

Characteristics that would be masked (not expressed) in generations of normal fish can be studied in the clones. In the picture on page 21, the adult fish at the top of the page is a normal zebra fish. The other fish are examples of clones that show variations in pigmentation, or coloring. Although just four different clones are shown here, the Oregon scientists have produced fifty-five different clones of zebra fish. In the other picture, embryos three days after fertilization already show the differences between normal fish (upper) and abnormally pigmented fish (lower).

Once established, a clone of fish can be maintained by normal sexual reproduction. Most of the offspring are females, but males can be created through the use of hormones.

Scientists believe that the techniques used to clone zebra fish can be applied to fish that are used as food. A combination of cloning and fish farming may eventually produce more fish such as trout, salmon, and catfish, thus helping to increase the world's food supply.

There is a vast difference between the cloning of frogs and fish in which the eggs are large enough to be seen and the cloning of mammals in which the eggs are microscopic and more fragile. Still, some scientists are trying to produce identical animals that will help in research in cancer, aging, birth defects, and the increased production of cattle for food, and in other areas.

Cloning by nuclear transplant has long been the goal of biologists, especially those who are concerned with cattle breeding. This technique involves the transfer of the nucleus of a body cell from an embryo into the egg cell of another organism. Work along these lines has been somewhat successful with animals such as fish and amphibia. For example, it has been possible to produce normal adult frogs with nuclei from tadpoles. However, a major milestone was reached when nuclear transplants succeeded in mice. This was done by removing nuclei from cells of mouse embryos and transplanting each nucleus into a single-cell mouse ovum, or egg cell. The ovum had already been activated by a fertilizing sperm, and the nucleus of this egg cell had been removed.

In June of 1983, researchers at the Wistar Institute of Anatomy and Biology in Philadelphia reported another milestone in genetics. They achieved a success rate of more than 90 percent in transplanting nuclei from one mouse embryo to the embryo of another mouse of a different breed and color. The new mice are not clones of the mother, but since the nuclei are extracted from cells of the same embryo and placed in different foster mothers, they can be considered, in the broad definition, clones of each other. Although clones of many embryos have been produced, making genetic copies of adult animals by transplanting nuclei from the body cells of adults into ova is another matter.

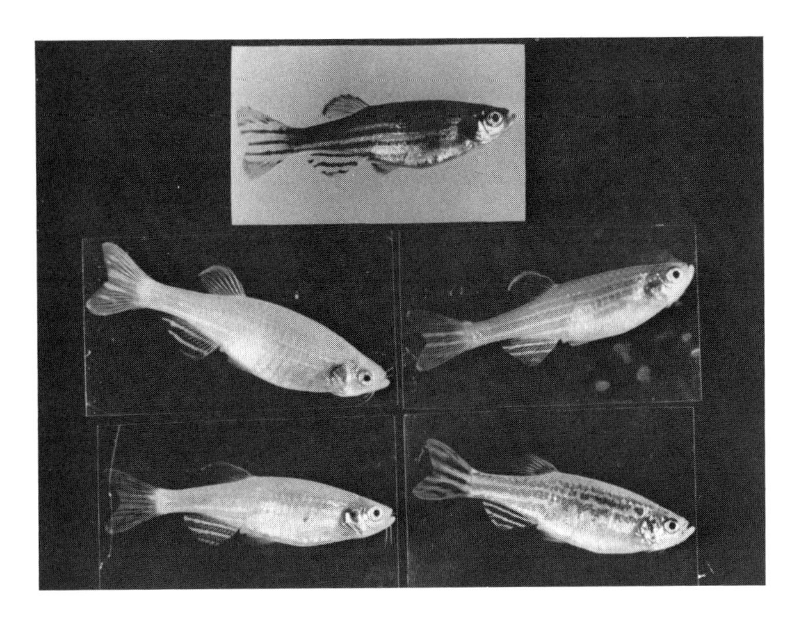

Figure 1-3. Normal adult zebra fish (top) about 3 cm long, and four of the many varieties with pigmentation mutations that have been cloned (bottom).
Photograph by Harry Howard, University of Oregon

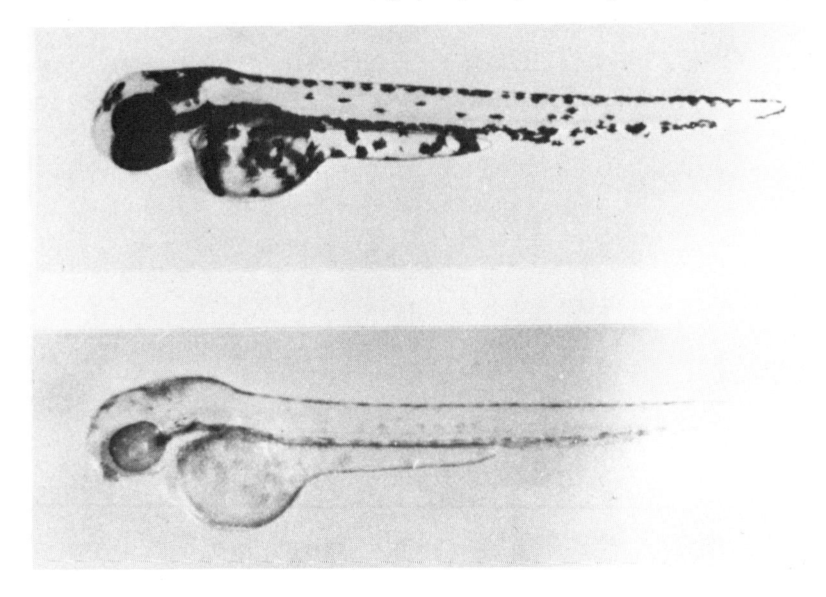

Figure 1-4. Three-day-old zebra fish embryos about 3.8 cm long already show pigment differences between normal (top) and mutant (bottom) fish.
Photograph by Harry Howard, University of Oregon

Many biologists are at work trying to produce special breeds of cattle by another cloning method. They are trying to perfect a cloning technique that will split a single embryo of thirty-two cells into two or more pieces. These pieces can be split further as they grow, and then they can be inserted into surrogate mothers where they will grow into cattle with the genetic makeup of the embryo. This would produce large numbers of clones similar to twinning and other multiple births in nature. To comply with the standard definition of cloning, the fertilized egg from which the cloned cattle originated would be considered the single parent.

When fertilized eggs or whole embryos are transferred from one cow, sheep, or person to another and the embryos grow in surrogate, or substitute, mothers, scientists agree that no cloning is involved. However, people often confuse this technique with cloning.

Should scientists tamper with the embryos of cattle? Will scientists someday tamper with human embryos to produce clones of human beings? Such questions cause a great deal of debate.

Cloning experiments with animals are being performed for a number of reasons, most of which are considered far more important than the purpose of producing carbon copies of individuals. For example, cloning plays a part in cancer research and in the production of diagnostic tools for medicine. One of the greatest mysteries in the world of cancer research is what makes a gene give directions to produce abnormal or cancerous cells. Cloning may help to solve this problem.

Before you reach a decision about whether or not you feel that cloning and genetic engineering should be continued under strict guidelines or even continued at all, you may wish to consider the applications of genetic engineering that are described in the following chapters.

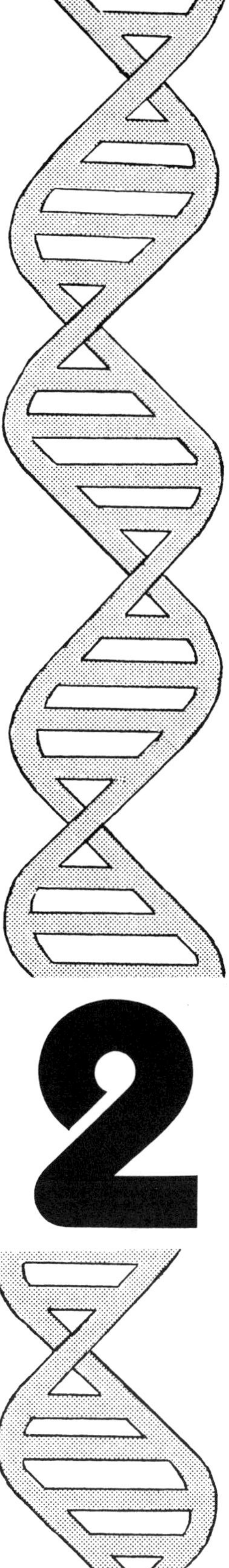

2

Your Genes and the Stuff of Life

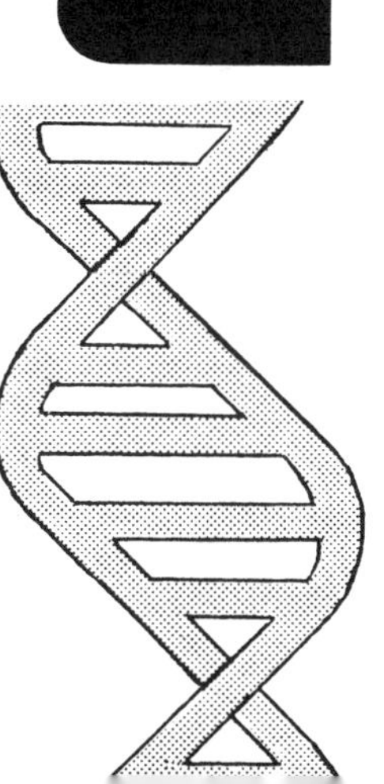

Genes are the basis of the whole science of genetics and the exciting revolution in biology. Although cloning is but a small part of that revolution, the role of genes is a major part of cloning and the applications of genetic engineering. Most people know about genes only because of the common expression, "Blame it on your genes."

Has anyone ever told you that your genes were responsible for your hair color, your good disposition, your quick temper, your shortness or tallness, or other characteristics? Some of these factors depend more on genes than others, but genes are praised and blamed for almost anything.

Genes are packets of chemicals that contain the codes for the development of living creatures. Understanding of the way genes direct the workings of the human body and that of other animals and plants is proceeding at an astonishing pace. Many words that are common today in the world of biologists did not exist thirty years ago. Today, scientists can make human genes grow in bacteria, manufacture genes in their laboratories, and produce new life forms by combining genes of different organisms.

Genes, the coded messengers, are present in every cell in all living things. Even a simple bacterium has about two thousand or three thousand genes that determine its life functions and structure. Each single yeast cell contains about ten thousand to twenty-five thousand genes.

In addition to the many thousands of genes that function in a myriad of ways in each human cell, there are copies of the same genes repeated again and again with minor alterations. This appears to be a system that allows the cells to produce needed materials under different sets of environmental conditions.

Genetic instructions in the nucleus of a cell determine what each cell's function will be. The cell's machinery reads the code contained in each gene and, through a complex process, converts nutrients into specific proteins. Each gene contains information for a single protein.

When assembled, proteins take part in the determination of eye color and creation of brains and all the organs of the body, as well as the massive communication network that determines how they function and keep the body healthy. Together the genes can be considered as a map, a guidebook, or a blueprint for a living thing. Genes determine the eventual appearance of a person, a rabbit, an orchid, or any other kind of living thing and whether the person is black, the rabbit is white, the orchid purple, and far more.

The part that genes play in the lives of human beings and all other animals and plants is important indeed. But while genes direct the expression of physical characteristics such as eye color, wings, fins, and feet, every organism is influenced in some way by the environment in which it develops.

Almost everyone agrees that both heredity and environment influence many physical characteristics, but the influence of genes is, and has long been, the subject of much controversy. Recent experiments seem to indicate that the way your genes express themselves plays a larger part in influencing your behavior and emotional life than previously believed. Hundreds of adopted children are being compared to both their adoptive parents and their biological parents. Large groups of identical twins, who have identical genes, are being studied too. Scientists are trying to determine the extent of hereditary-factor involvement in shyness, mental illness, stuttering, alcoholism, and certain learning disabilities, and other areas. While the results indicate that you may blame some of your unwanted characteristics, such as shyness, partly on your genes, there is still much to be learned.

Genes have been called the center of the revolution in biology. The basis of the new biology was laid down many years ago when an Austrian monk studied the sweet pea plants in his garden. Gregor Johann Mendel combined his two hobbies of mathematics and botany, and began his serious experiments in 1856. Through eight years of meticulous observations, he formulated the basic laws of heredity. Mendel noted that specific characteristics of his sweet pea plants, such as height and color of blossom, were passed on to new plants from parent plants. Mendel assumed the appearance of a characteristic was controlled by a pair of factors, one factor from each parent. Today, we call these factors genes. Before that time, ideas about heredity were based on speculation and superstition.

Mendel showed that genes were transmitted as complete units that retained their individual characteristics and that they were transmitted in a specific way. Today, the laws of heredity thai illustrate the ratio of certain dominant to recessive, or expressed to silent, traits are familiar to all biology students, but Mendel's work was not recognized as a great achievement during his lifetime. The importance of these carefully recorded experiments, published in 1866 by a local nature society, was not really appreciated until the beginning of the twentieth century when scientists learned more about the structure of living things and how they reproduced.

Doctors and other scientists who have long been treating many diseases as totally due to either heredity or environment are slowly changing their approach. The nature of all illnesses is complicated, but the new way of viewing all disease takes into account both heredity and environment. In some diseases, environmental influences appear to play a larger part, while in others, the genes have the major role. For example, babies who are born with PKU (phenylketonuria) are deficient in a certain enzyme. The absence of this enzyme prevents the body from making use of an amino acid that is found in most protein-rich foods, and its absence almost always leads to mental retardation and other problems. PKU would appear to be a purely genetic disease, but it can be controlled if recognized early enough. Many states have made the PKU test mandatory for all newborn infants. By giving an infant with PKU a carefully regulated diet as early as the first few weeks of life, thus altering the infant's environment, this disease can be brought under control.

The most common ailment of the circulatory system—high blood pressure, or hypertension—has often been blamed on too much stress, overwork, and other environmental conditions, but researchers have found that there is a strong genetic factor involved.

One of the approaches to the study of the part played by genes versus the part played by environment is opposed by many theologians, geneticists, sociologists, and others. This approach is the use of sperm from men of very high intelligence to father babies in an effort to produce very bright children. Many serious scientists object to the shadow this kind of experimentation throws on work aimed toward the healing of diseases, the production of more food, and the elimination of pollutants.

Genetic engineering is the technology that is used at the laboratory level to alter the hereditary apparatus of a living cell so that the cell can produce new chemicals or perform new functions.

Some individuals consider efforts to alter defective genes as a means for disease prevention to be a dangerous procedure; many others believe that one of the exciting aspects of genetic engineering is the possibility of controlling these defective genes. Estimates indicate that every person may carry from four to eight genes which, when combined with the same defective genes from a mate, can produce offspring with a genetic disease. But such unfavorable combinations are relatively rare because, in many cases, a defective gene is "silent" when combined with a normal

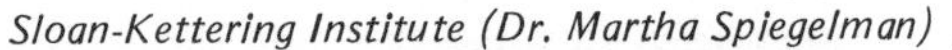

Figure 2-1. A normal mouse embryo of 16 cells. Differentiation has not yet begun, but the cells have compacted as if they had agreed to cooperate as a single organism.

Sloan-Kettering Institute (Dr. Martha Spiegelman)

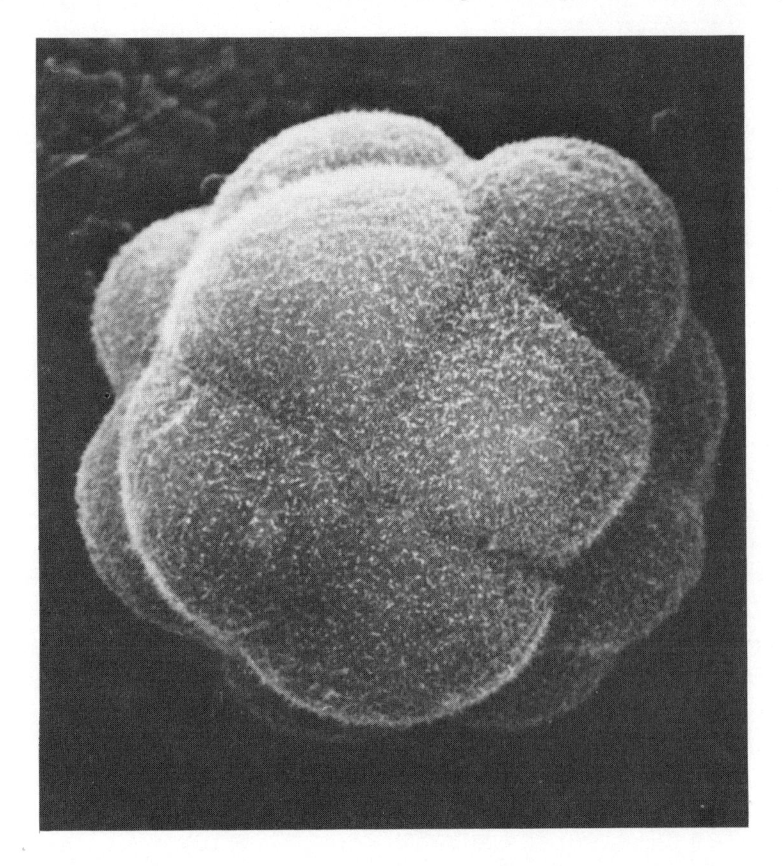

gene for the characteristic that is contributed by the other parent. Genetic counseling helps some parents discover whether or not both carry the same defective genes.

According to the Department of Health and Human Services, over 15 million Americans suffer from one or more types of birth defects, 80 percent of which are thought to be caused by genetic changes. Mankind is heir to some 3000 different genetic diseases, but fortunately, most of them are very rare. However, the life-years lost to genetic diseases are estimated to be six and one-half times as many as those lost to heart disease. Obviously, such diseases are important.

Figure 2-2. A mouse embryo of 16 cells with a lethal genetic defect located near the genes that control tissue compatability. Instead of compacting, these cells are growing independently and will not differentiate.

Sloan-Kettering Institute (Dr. Martha Spiegelman)

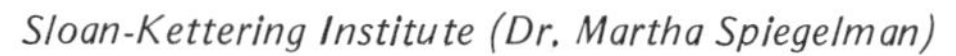

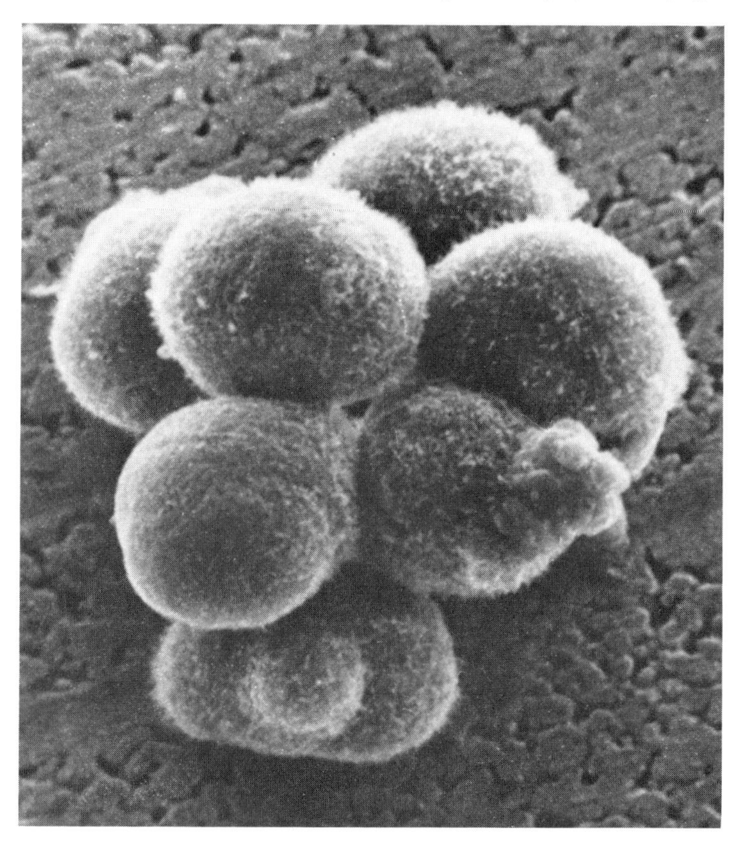

No one actually knows the degree to which genes are involved in the hundred different kinds of cancer, but in recent years, scientists have identified certain genes that are responsible for some kinds of the disease. They know that a change in 1/6000 of the chemical makeup of a gene can alter the behavior of the cell and breed cancer. Scientists believe that the actual change is provoked by some environmental effect on the body, such as stress, chemical pollution, or smoking. They hope to eventually learn exactly what makes genes trigger cancer and to find ways to prevent it. This is just one of the goals of genetic engineering that is widely accepted.

How are genes related to DNA? How can scientists manipulate genes to correct the defects found in nature? How can clones, cells that are genetically identical, help to keep people healthy? These and other important questions will be answered in Chapter 3.

Figure 2-3. Mouse leukemia is triggered by a viral cancer gene. The loops of a single-stranded DNA where the specific gene is located are shown (above) in an electron micrograph (magnified more than 300,000 times). A schematic drawing (below) clarifies the double and single strands.

Sloan-Kettering Institute (Dr. Allen Oliff)

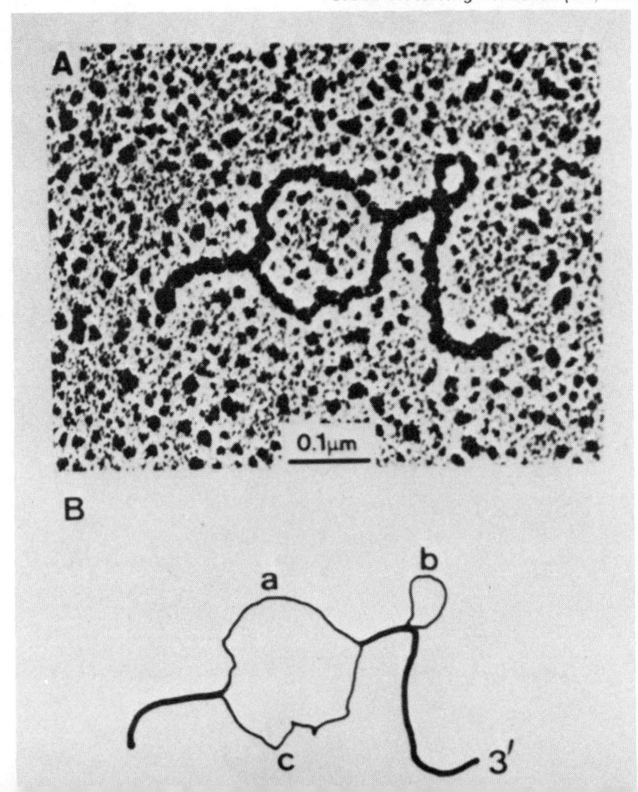

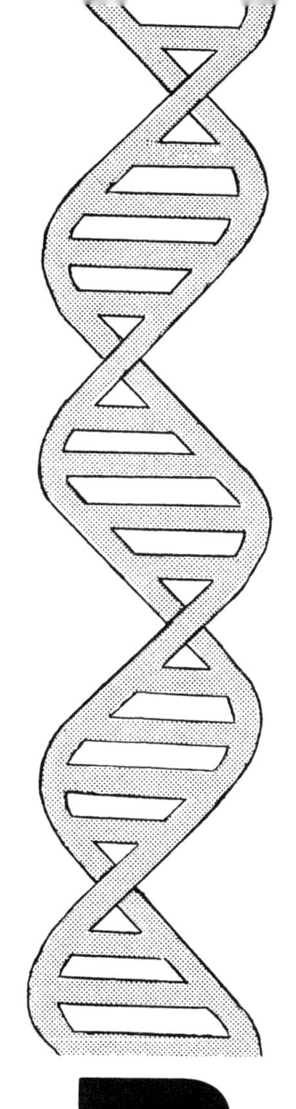

3

DNA:
The Code of Life

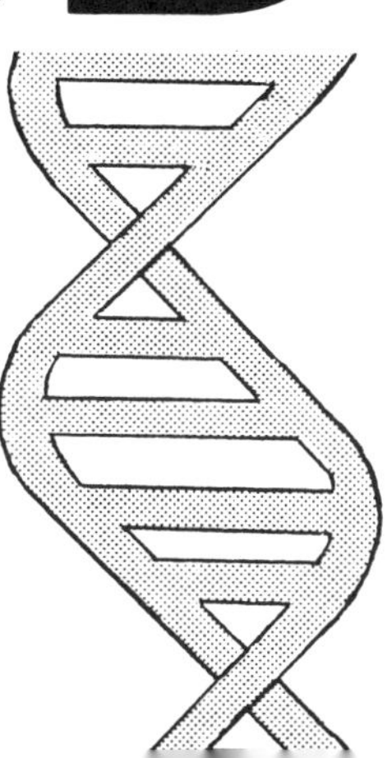

Almost everyone has heard of DNA but few people can tell you what it is and how it is related to genes. Actually, genes are linear stretches of DNA (deoxyribonucleic acid), the universal material that governs heredity in all of the millions of different species of living things. How did scientists begin to solve the puzzle of heredity?

Genes are found in the nucleus of a cell in microscopic filaments, or strands, called chromosomes. The material of the chromosomes was seen more than a hundred years ago by the German chemist Friedrich Miescher, who soaked bits of different plant and animal tissues in an assortment of acids. Miescher suspected that the insoluble dark material that was left had something to do with heredity. Actually, he had purified DNA.

Each chromosome is made of a tightly coiled strand, actually a single molecule of DNA. A chromosome duplicates itself, forming two DNA molecules, in preparation for cell reproduction. About 99 percent of the DNA in a cell is found in the chromosomes, along with other material such as protein.

DNA has been called the code of life, and it has also been called the material link between generations. Recent genetic research at the University of California at Berkeley has uncovered a human interest story. Researchers found that at least nine inbred strains of laboratory mice have the same ancestor, the very same "grandmother." They identified the DNA in cell structures called organelles, or mitochondria, and this DNA is maternally inherited. What they found set the nine strains distinctly apart from others and indicated that they were all descendents of one mouse. This "greatest grandmother of all" may now have billions of grandchildren among the mice that are being used in laboratories to study aging, cancer, blood disorders, and in numerous other kinds of medical and biological studies. The dowager mouse, the first of these nine strains, probably lived in the first part of the twentieth century, although it is not impossible that she lived as long as 3000 years ago, when mice were domesticated in China.

Even before scientists recognized the importance of the chemical DNA, they learned from the work of Thomas Hunt Morgan that genes were not completely independent units as Mendel had believed. Working with fruit flies, Morgan also showed,

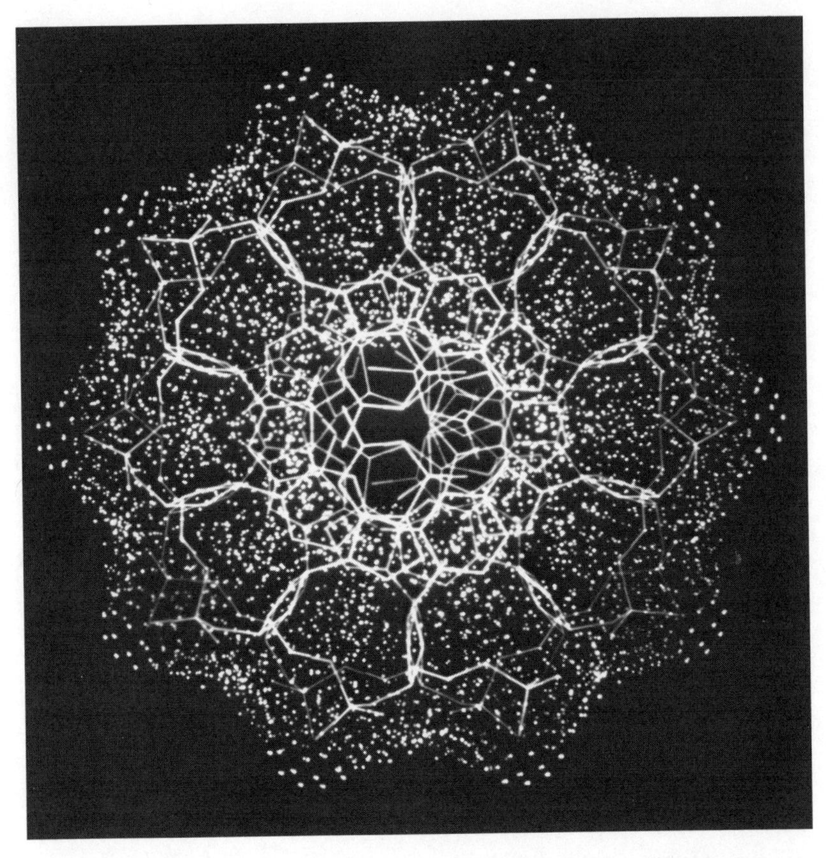

Figure 3-1. DNA as seen along the axis of the double helix. This and Figure 3-2 are computer-reconstructed molecules.

Computer Graphics Laboratory, UCSF

---

in the 1920s, that genes were not always permanent; they could change. Natural changes, or mutations, occur at a very slow rate, but exposure to X-rays greatly increases the rate of change.

In the 1940s, new observations helped scientists to discover what the genes were made of. After meticulous research, a group of scientists showed that the basic structural unit of heredity is DNA, a large molecule that is built from a large number of very similar chemical building blocks. Many scientists found it difficult to believe that a molecule with so much monotonous repetition could be responsible for the vast number of products and reactions in living things. They wondered how it could faithfully reproduce itself so that information could be transferred from generation to generation.

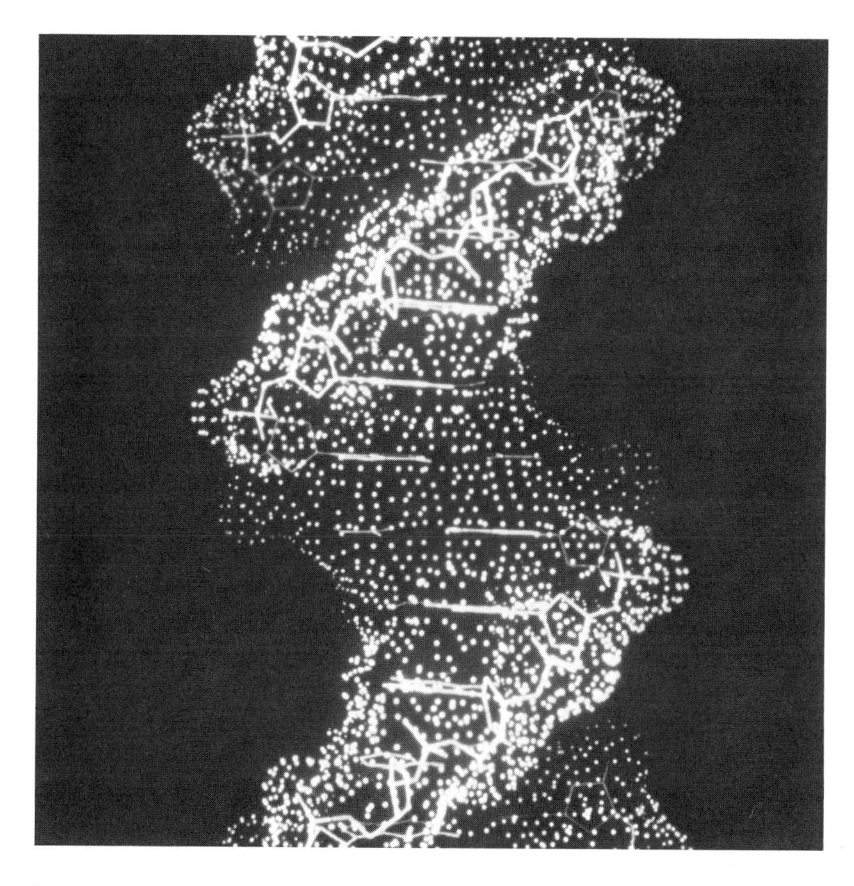

Figure 3-2. DNA as seen from the side of the double helix.
Computer Graphics Laboratory, UCSF

The ingenious work of James Watson and Francis Crick and others in building a model of the DNA molecule removed any doubts that scientists had about this puzzling molecule. Without ever seeing the molecule itself, Watson and Crick used chemical analysis and X-ray diffraction studies completed by other scientists to construct a model that showed how parts of DNA were arranged. The actual structure of DNA was too small to be seen by the electron microscopes of 1953 when their model suggested the way the chemicals in DNA were joined, but it was photographed by a powerful electron microscope in 1969. Today, the long, ribbonlike molecule which is twisted around itself and called the double helix is familiar to many.

DNA is often described as a graceful, spiral ladder. The sides of the ladder are alternately sugar and phosphate molecules, and the rungs are chemical bases known as purines and pyrimidines. The two purines in the rungs are adenine and guanine, represented by **A** and **G**. The two pyrimidines are thymine and cytosine, represented by **T** and **C**. These four bases form the four letters of the alphabet of the genetic code. Just as English consists of a twenty-six letter alphabet and a computer reads a two-letter alphabet, the genetic language has four letters, or codes. All genetic messages in all forms of life are written in this four-letter alphabet.

In the DNA from any kind of plant or animal, **A** (adenine) pairs with **T** (thymine) and **G** (guanine) pairs with **C** (cytosine). No matter whether the source of DNA is a microbe, a mouse, a monkey, or a man, it contains amounts of **A** that are equal to the amounts of **T** and amounts of **G** that are equal to the amounts of **C**. And since DNA is the master blueprint, or universal language, of life, almost all living things contain DNA. Viruses are an exception. Not everyone considers viruses as living things. A virus is not a cell. It is essentially a piece of DNA or RNA with a coat of protein.

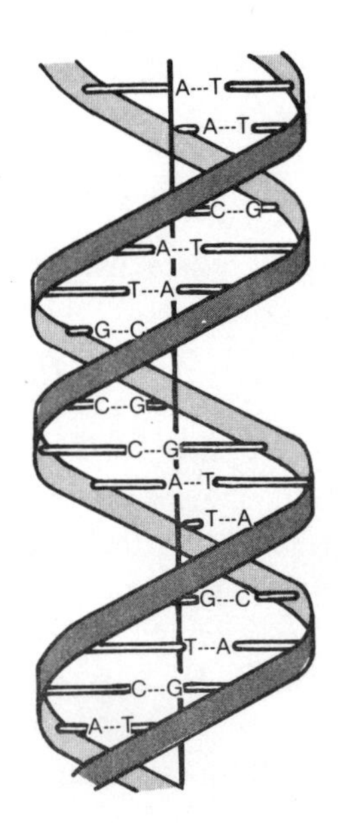

Figure 3-3. A schematic diagram of the DNA double helix.
Office of Technology Assessment

How can so much information be stored in molecules with so few different kinds of material? Even with only four different chemicals in the rungs of the DNA ladder, there are almost an endless number of different combinations. It is the variation in the sequence of these bases along the sugar-phosphate backbone that contains the code later expressed as proteins. The sugar-phosphate chains twist around the outside, and the paired bases hold the two chains together. All DNA, whether from a bacterium, a boxwood shrub, or a bear, is made of the same chemicals. Only the sequence of the base pairs differs. When you read a magazine, the articles have meaning for you because of the order of the letters. In plants and animals, the order of the four bases gives meaning to the genetic material.

A DNA molecule is tremendous in size when compared with the size of other molecules. Even though DNA is a large molecule, molecules themselves are so small that it is difficult to comprehend how small DNA really is. Paul Berg, a Nobel prize winner for his work in genetics, has suggested that the DNA coiled in all the cells of a single human being would cover a thread nearly 500 million miles long if it were stretched out in a single line. This is more than five times the distance between the earth and the sun.

Figure 3-4. An electron micrograph showing a plasmid, a tiny DNA molecule.
USDA Agricultural Research Service

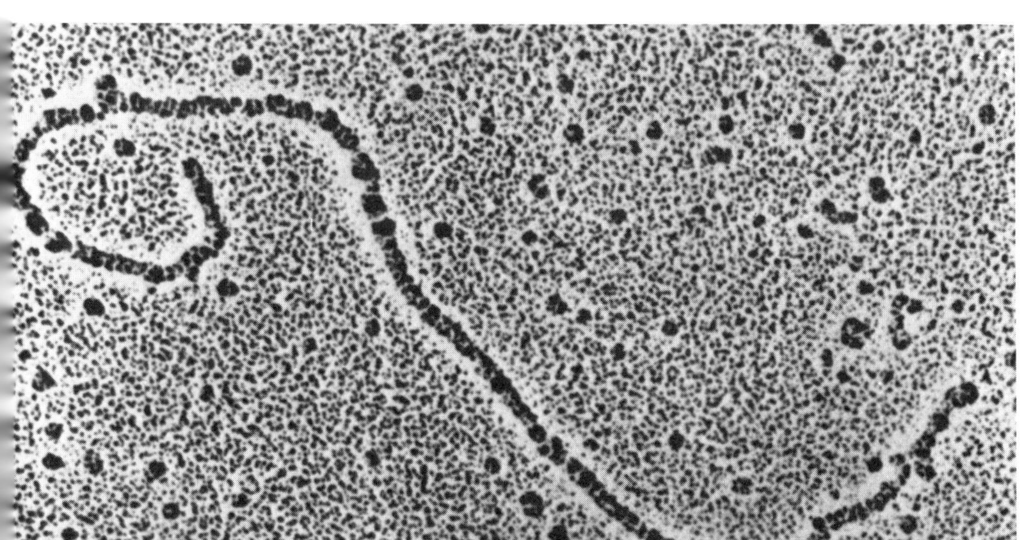

It is not surprising to find that the DNA of an animal such as a dog has far more rungs in its ladder than a simple species such as a bacterium. A virus may carry its genetic instructions with DNA that has about six thousand rungs, while a complex animal may have DNA that contains many millions of base pairs in its ladder. Each cell in the human body, with the exception of red blood cells, carries the DNA blueprint.

DNA is copied for two reasons: to provide the information for constructing new proteins and to perpetuate the species. One of the most exciting things about DNA is the way in which it makes a copy of itself. Each new body cell that forms receives a perfect copy of its DNA. The DNA molecule has been compared to a zipper, with the base pairs as the teeth and the sugar-phosphate chains as the strands of cloth to which each zipper half is sewn. Thinking of DNA as a zipper helps scientists to understand how the molecule can duplicate itself when cells divide and how a gene, a section of the DNA, can be passed in exact duplication from one generation to the next. The DNA double helix molecule "unzips" when dividing, and each half then forms a new molecule.

One half of the DNA in your body came from your mother and one half from your father. Each human cell has twenty-three pairs of chromosomes, forty-six chromosomes in all. Certain cells divide to make egg cells or sperm cells, each containing a single set of twenty-three chromosomes. When an egg cell is fertilized, the chromosomes from the sperm are paired with the egg's chromosomes and there are forty-six chromosomes again. In the process of the manufacture of eggs and sperm, there is some shuffling of genes, so that any one chromosome in a sperm, for example, could be a patchwork of genes from the mother and father. No two sperm cells are identical, and no two egg cells are identical. As a result, brothers and sisters have many common genes, but their genes are not identical unless the siblings are identical twins. Identical twins are formed when one egg splits and separates, giving rise to two individuals. Identical twins are always the same sex. Clones would be identical in their genetic makeup; the question and problems of human cloning will be discussed later.

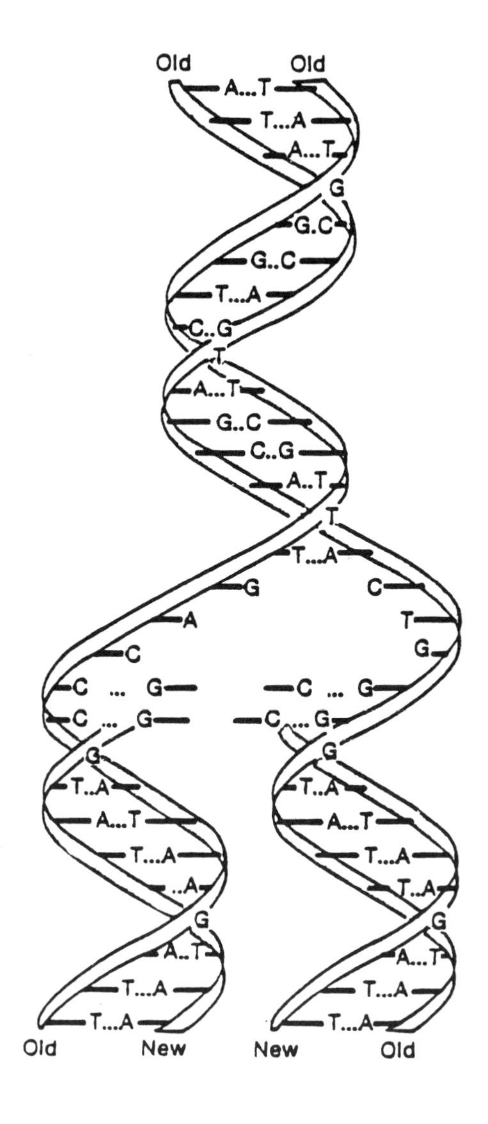

Figure 3-5. A simple interpretation of DNA making a copy of itself.

Office of Technology Assessment

Genes are the functional units of DNA, the sequences of chemicals that are strung along the DNA molecule. Each gene, or sequence, contains the information for—codes for—a specific protein. But segments of DNA, some of which have no known function, appear at the ends of the genes, and even the genes themselves have sections that are not part of their coding instructions. Scientists are still exploring these noncoding regions of DNA, but they do know a great deal about the chemical composition of DNA, which is often referred to as the "stuff of life." DNA in the genes is transcribed into a chemical known as messenger RNA (ribonucleic acid), which is then translated by reactions in the cell into protein.

In spite of the tremendous speed of advances in all aspects of knowledge in the field of genetics, so much remains to be learned that many scientists predict exciting discoveries for the future.

4 The Splice of Life and Gene Cloning

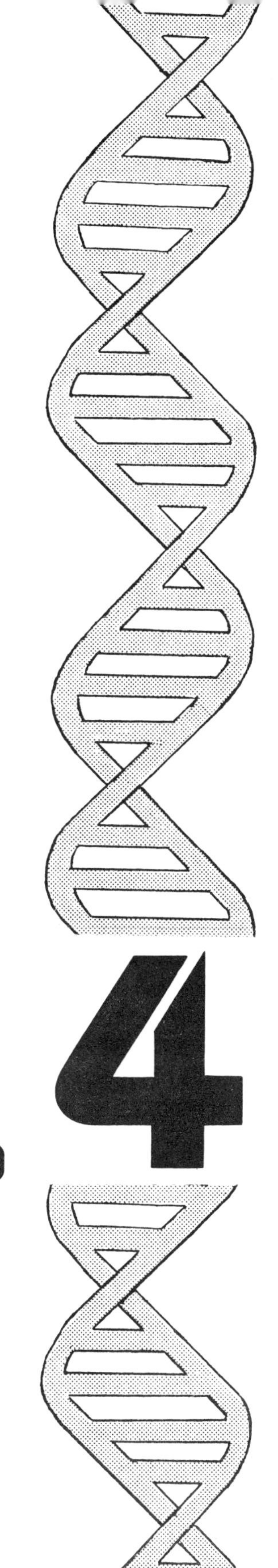

Making genes? Impossible! But the making of gene fragments has moved from the realm of science fiction to the real world of science. Biologists can buy gene machines for their laboratories that look somewhat like a set of boxes on a desk top. These gene synthesizers are stocked with the chemicals needed to make subunits of genes that can be combined in millions of different ways. The linking of chemicals that once took many months of tedious work by scientists in their laboratories can be accomplished in a fraction of the time with gene machines.

When scientists need a special gene segment for research, they punch the exact code with the building blocks in proper order into the keyboard of the computer. After a period of time that depends on the length of the chain that has been ordered, the machine delivers in precise sequence the exact amount of the desired chemicals needed to synthesize the desired gene. Synthetic gene fragments, with as many as twenty units chained together, or entire genes, are released into a vessel of the machine. Whole genes can be made from these subunits by chemically fusing the necessary segments.

Figure 4-1. A typical gene machine used to synthesize genes.
Genex Corporation

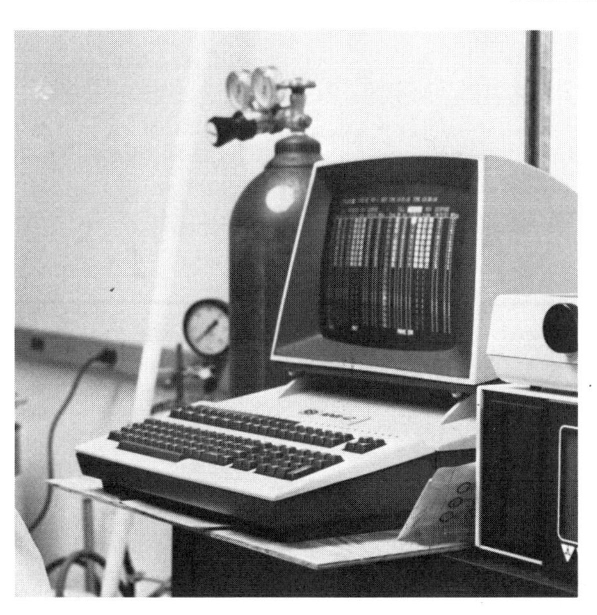

The idea of gene machines may sound fairly simple to people who have no idea of the long trail of experimental work that made them possible. A very large number of scientists contributed to the knowledge of what makes a gene.

Gene splicing is a method in which pieces of genes or whole genes are inserted into bacteria. The bacteria multiply and make large amounts of the protein called for by the newly inserted gene. This gene could be derived from the use of the gene machine or as a result of a process designed to isolate a special gene. Such a gene could code for insulin or antibiotic production. Thus gene splicing can produce medically important substances.

Not too long ago, there was great excitement when an announcement was made that a 79-year-old grandmother was the first person to be treated with interferon in part of the largest clinical study to be conducted so far. She had cancer of the lymph nodes and had been accepted as part of the study because other treatments, such as radiation, surgery, and chemotherapy, had all failed. Interferon is one of the wonder molecules of recent years. It is a natural protein that can now be produced in quantity in laboratories by genetically engineered bacteria. Doctors hoped that the intramuscular injections would help this grandmother by attacking her cancer cells and activating her body's own immune system.

In the same year, it was announced that Humulin, a genetically engineered form of insulin, would be available in local pharmacies. In the past, diabetics who were allergic to the common form of insulin, which is derived from animal tissues, had to try to control their diabetes through diet, weight control, and the use of other drugs that have unpleasant side effects. Now, genetic engineering has made it possible for large amounts of synthetic insulin to be produced in the laboratories. This insulin is identical to that produced in the human body. Humulin and interferon are discussed more fully in the following chapter.

These are just two illustrations of practical applications of the new genetics. Genetic manipulation is moving out of the laboratories to benefit an increasing number of people. The above true cases happened in 1982, a year in which long periods of hard work seemed to produce practical applications of the genetic dream at a faster pace.

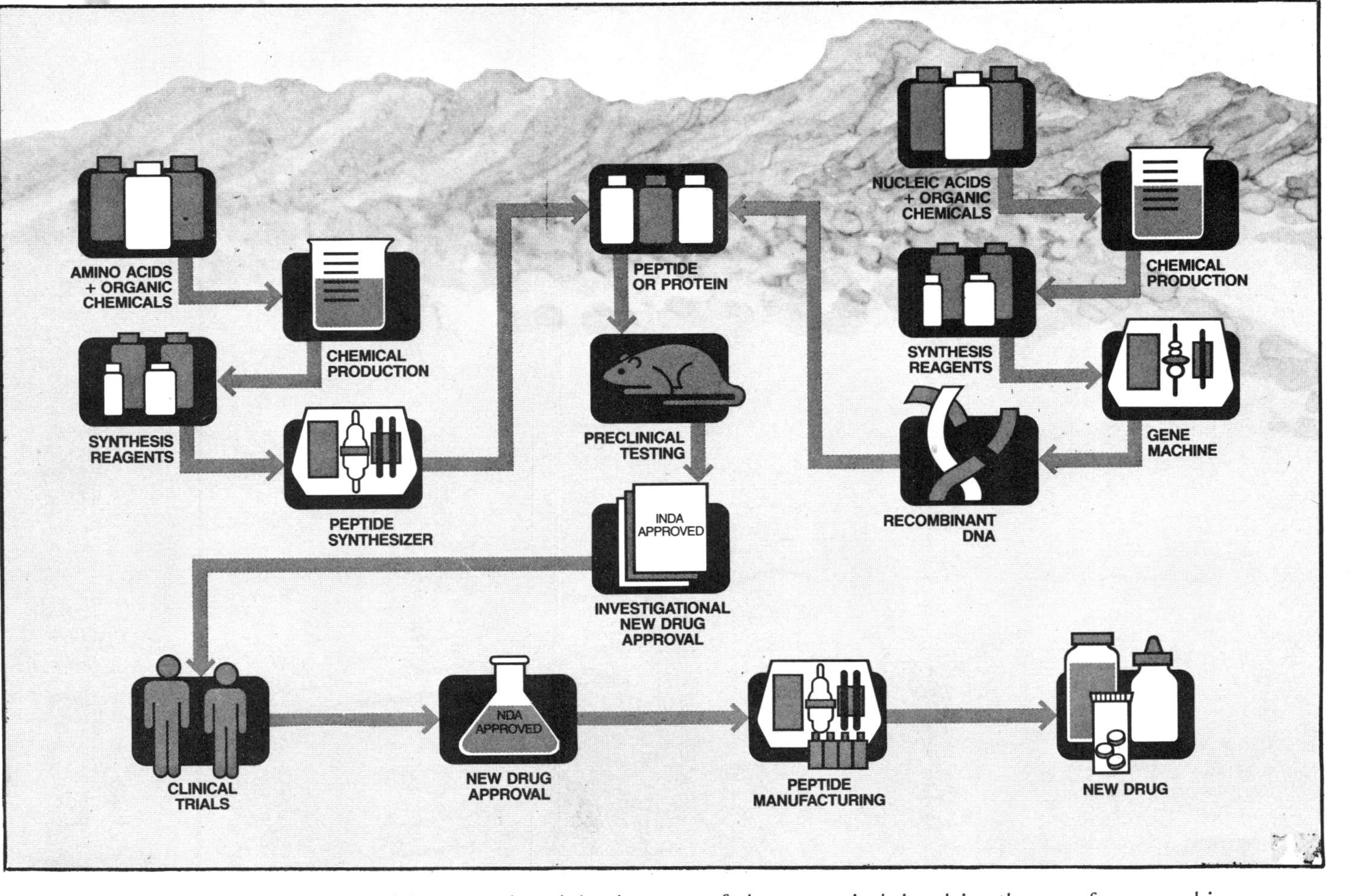

Figure 4-2. A simplified diagram of the research and development of pharmaceuticals involving the use of gene machines.

Vega Biotechnologies, Inc.

Few people who read about the practical progress in genetic engineering have any idea of the tremendous amount of work or the true measure of such accomplishments. The splicing of life that made them possible is based on an unbelievable amount of serious experimentation with materials that are too small to imagine.

Dr. David Jackson, one of the pioneer gene splicers, has suggested that people think of a DNA molecule as a million times larger than it really is. Then it would be as thick as a piece of string about one-eighth inch in diameter. On this scale, a typical gene would be about twenty inches long. Now think about the fact that a simple bacterium has enough DNA for 3000 to 4000 genes. This amount of DNA would be just over a mile of string, and the bacterium itself would be the size of a box three feet wide, six feet long, and three feet high.

Dr. Jackson makes the small size of the materials with which the geneticists work even more real by the following illustration: Using a scale which increases the size one million times, the amount of DNA in a cell of the human body is equivalent to 1000 miles on the string. He suggests that one can get a better idea of the genetic complexity of the DNA in a mammal by imagining the string stretched between New York City and St. Louis. If the number of genes involved in this imaginary DNA were placed along it at even intervals, you would see a gene every two feet of the way.

In spite of the unbelievably small size and large complexity of genes and DNA, new developments in the field of genetic engineering appear rapidly. Some techniques that were discovered about a decade ago are responsible for some of this rapid development, but these techniques were based on research that began many years before, thus providing a gradual increase in the understanding of the composition and function of genes.

For generations after the discovery of genes, it appeared to be a fundamental truth that genes from one animal or plant could not enter the hereditary apparatus of a different one. Today, it is recognized that this is not the case. However, there was great excitement when it was first learned that viruses can alter the hereditary material of the bacterial cells that they infect.

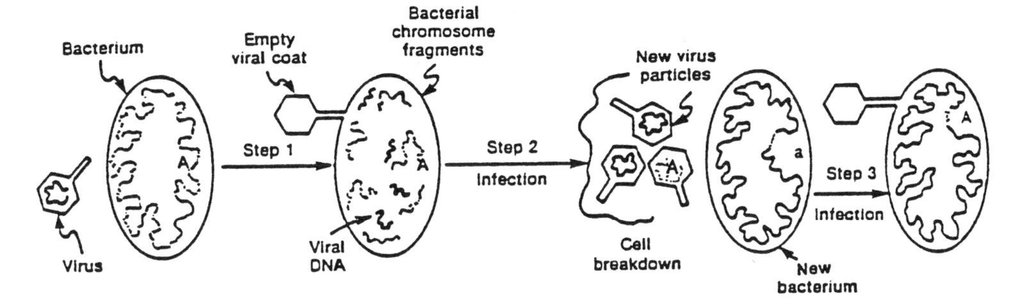

Figure 4-3. The transfer of genetic material in bacteria by means of viruses.
Office of Technology Assessment

A virus is made of a DNA or an RNA molecule surrounded by a protein coat that protects it. Viruses have been called naked genes, hereditary material on the loose, and many other names. They are well known for the diseases they cause in humans, such as smallpox, influenza, mumps, measles, colds, yellow fever, and poliomyelitis.

Viruses cannot reproduce without the help of an appropriate cell of another organism. Some viruses attach themselves to their hosts and inject their DNA through the cell walls of their host plant. Bacteria are constantly under attack by viruses known as bacteriophage, or phage, that reproduce inside them. Some bacteriophage take over the entire cellular machinery of a bacterial cell and produce as many as 300 copies of the original virus. In other cases the viral DNA attaches itself to the bacterial DNA where it may become an actual part of the bacterial chromosome.

When the bacterial cell reproduces itself, it duplicates the attached DNA of the virus as well as its own. The original bacterium becomes the first of a new strain, a hybrid bacterium, in which the viral material is reproduced in true fashion. This genetic exchange from virus to bacterium is a natural kind of gene manipulation which enables a virus to reproduce. Thus, the viral DNA takes over the machinery of the bacterium's reproduction, using the bacterial DNA to make viral parts. The viral-bacterial DNA reproduces rapidly, and in the process there is an explosion of the bacteria, releasing numerous new viruses. The large numbers of viruses that are released travel off and attach themselves to new hosts.

Research scientists are especially interested in how the genetic material of the virus may be switched off, then suddenly be turned on again, making large numbers of offspring. Finding a chemical that triggers certain genes has long been a goal of research.

One of the popular organisms in which scientists study DNA is a species of bacteria, *Escherichia coli*, known commonly as *E. coli*. Many strains of *E. coli* are common; in fact, a harmless variety dwells in the colon of your digestive system.

A breakthrough in the science of genetics came early in the 1970s when scientists who were studying *E. coli* made an exciting discovery. They learned how to isolate a specific DNA sequence, or a gene, from one species and recombine it with a DNA sequence of another species. This technique produces recombinant DNA, a term that seems less difficult if one remembers that it means recombining DNA. The first scientist to construct a recombinant molecule, Paul Berg of Stanford University, opened a new era. He was awarded a Nobel prize in 1980 for this outstanding work.

One method of recombining DNA involves the use of small circular bits of DNA that are found in some bacteria, such as *E. coli.* Each *E. coli* bacterium possesses only one chromosome that codes for all cell functions, while it may also contain several ring-shaped pieces of DNA known as plasmids. Plasmids are the vehicles by which some bacteria inject some of their DNA directly into other bacteria when they mate in a process known as conjugation.

Figure 4-4. The transfer of genetic material in bacteria by conjugation.

Office of Technology Assessment

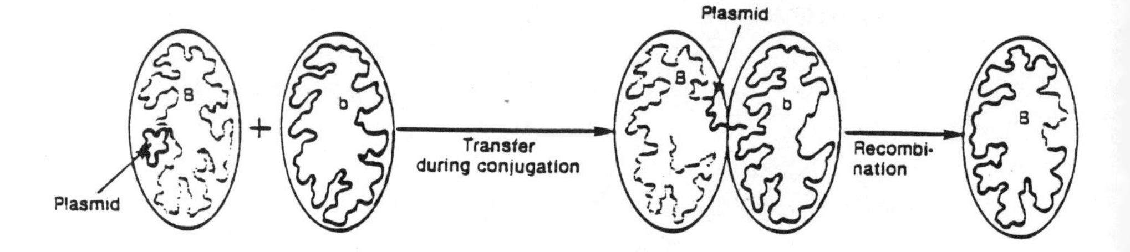

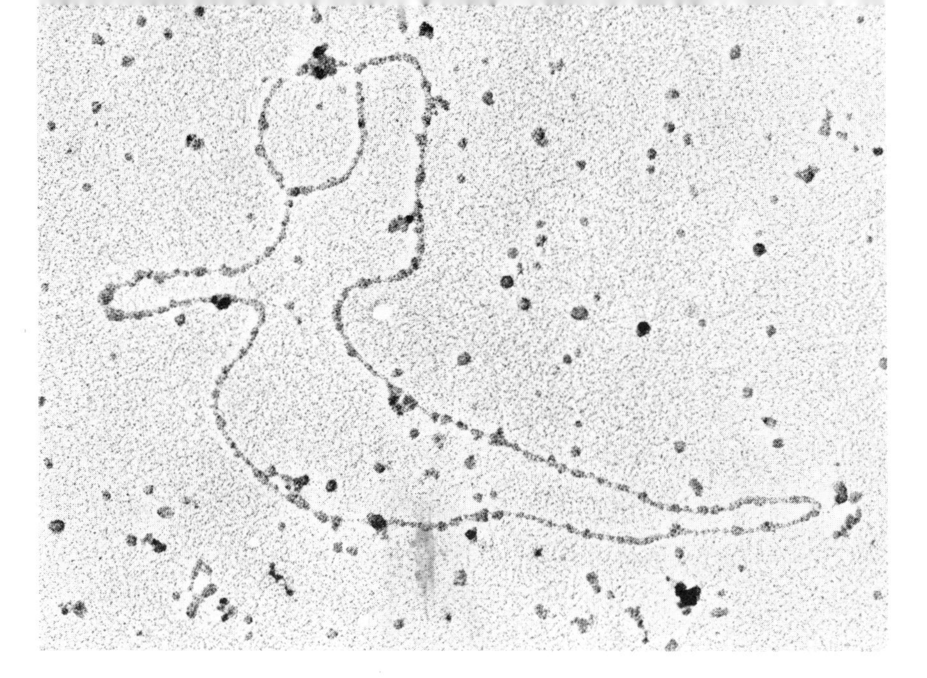

Figure 4-5. An electron micrograph of a replicating plasmid molecule.
Stanford Medical Center News Bureau

Apparently in an effort to protect themselves from viruses, bacteria have produced a number of enzymes called restriction enzymes. These enzymes snip out foreign pieces of DNA, making them useless. The restriction enzymes recognize specific areas of DNA that do not belong to the bacteria's own DNA and remove them, acting somewhat like a microscopic scissors. Each restriction enzyme recognizes one and only one specific sequence of DNA.

Scientists have identified a number of restriction enzymes and they use them in laboratories routinely. They can isolate a plasmid in *E. coli* or other organism and use restriction enzymes to cut the circular plasmid to make it linear. Again using restriction enzymes, they then isolate a piece of DNA from another organism that contains a specific gene or genes. They insert this second piece of DNA between the cut ends of the plasmid DNA and fuse the recombined plasmid into a circle again. The linking is made possible because a clean cut is not made by the restriction enzymes; they snip one strand a slight distance away from the other, leaving what chemists call "sticky" ends. This hybrid plasmid is then put back into a bacterial cell. The whole process can be compared to splicing movie film or tapes for electronic equipment. The added piece of DNA will be reproduced as part of the plasmid when it reproduces through its normal process. This is a way of making recombinant DNA, a way of recombining pieces of DNA.

Bacteria, usually *E. coli* strain K-12, are used as a host for the propagation of genes that scientists want. The plasmids are carriers that are used to insert the new DNA molecule into the host. Bacteriophages and yeast are sometimes used as carriers, too.

Some people visualize scientists using microscopic scissors to snip pieces of DNA from plasmids and microscopic tweezers for placing the genes into the DNA when they are making recombinant DNA. Of course, the extraction is made chemically in large batches of cells. Bacteria may be disintegrated chemically through the use of a detergent, then the plasmids can be isolated by the use of a centrifuge. The restriction enzyme cuts out the specific part for which it was introduced. Now there is a flask of the snippits of DNA and of open plasmid rings.

The pieces of DNA from human cells that contain the gene or genes that are to be introduced into the plasmids are added to the flask, and the two are mixed and combined with the help of certain chemicals. The pieces of DNA from the human cells close the circles of plasmids, a process that is helped by the fact that the ends that are open are sticky. Next, the plasmids are mixed with a new batch of *E. coli* or some other organism, and the plasmids are absorbed. Then scientists identify and isolate the colonies of bacteria that contain the desired DNA sequence and make large quantities of them.

Although the technique of recombining DNA may sound complicated, it is actually much more intricate than the description given here. The process of isolating, selecting, and reproducing a gene is commonly called cloning a gene.

Recombining DNA is not really a new process. It would be difficult to find in nature a DNA molecule that is not already a recombinant molecule. Some examples of natural genetic engineering in bacteria were mentioned earlier in this chapter. Bits of DNA change position naturally in the chromosomes of

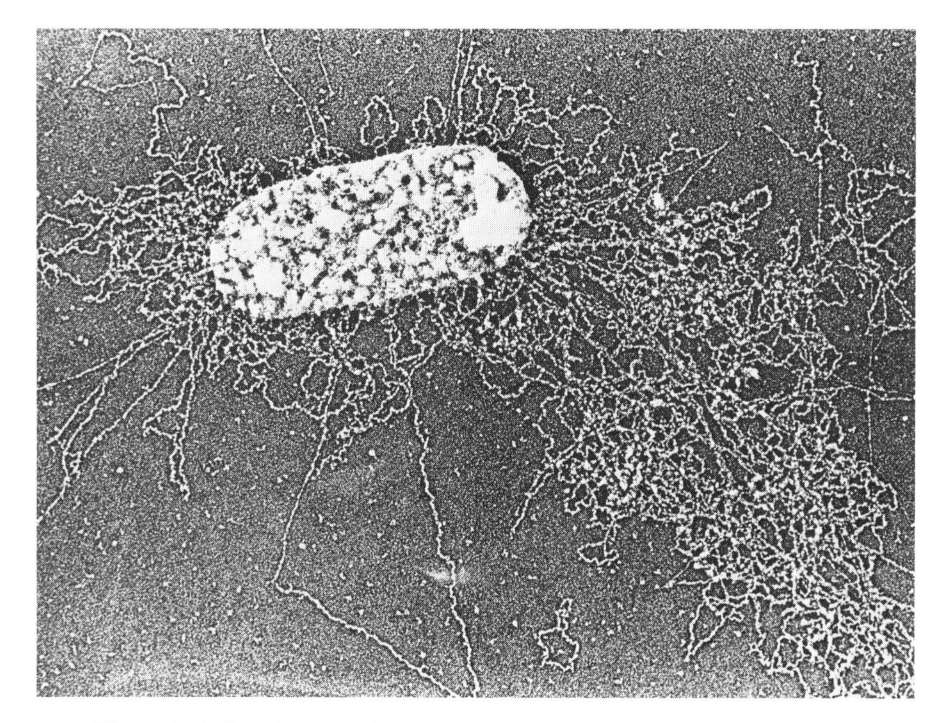

Figure 4-6. This electron micrograph shows *E. coli* releasing its DNA througl. a cell wall that has been disintegrated by a detergent.

Stanford Medical Center News Bureau

Figure 4-7. A chart showing how recombinant DNA is spliced and cloned.

Genex Corporation

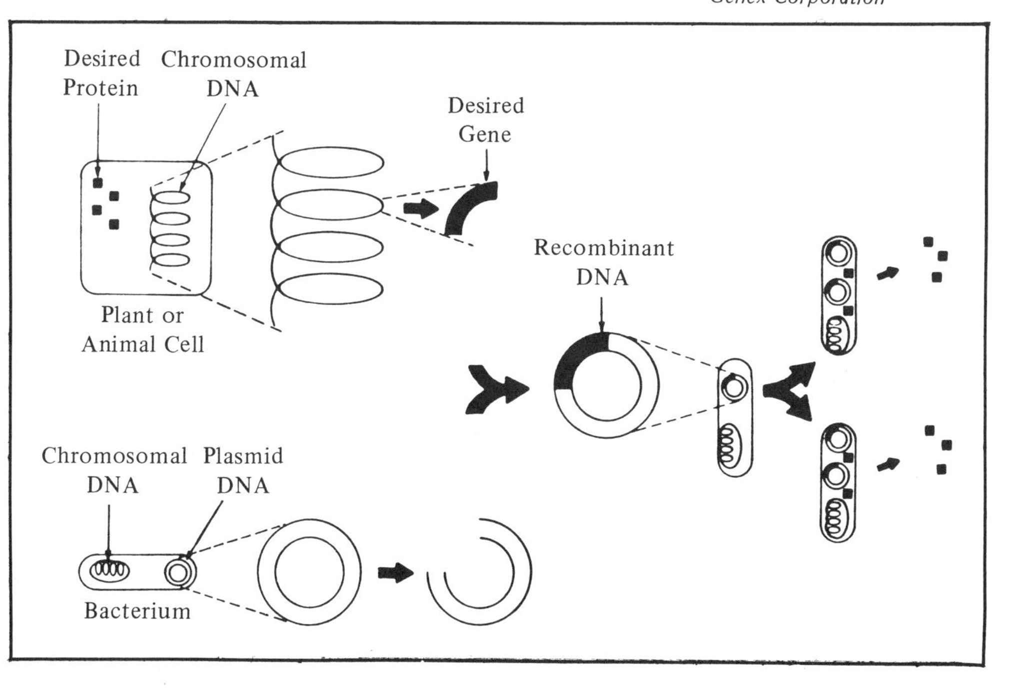

plants and animals. In fact, they have been doing so for centuries. However, this usually occurs within the same species and at random. Humans have been manipulating genes since the beginning of history through selective breeding of plants and animals.

It is the micromanipulation by humans with new speed and accuracy that is new. It is this splicing and cloning of genes that some people see as an event that will change the very foundations of agriculture, energy production, medicine, and the drug and chemical industries. Both the hope for great improvement in the quality of life and the concern about possible risk make progress in genetic engineering a subject of major importance.

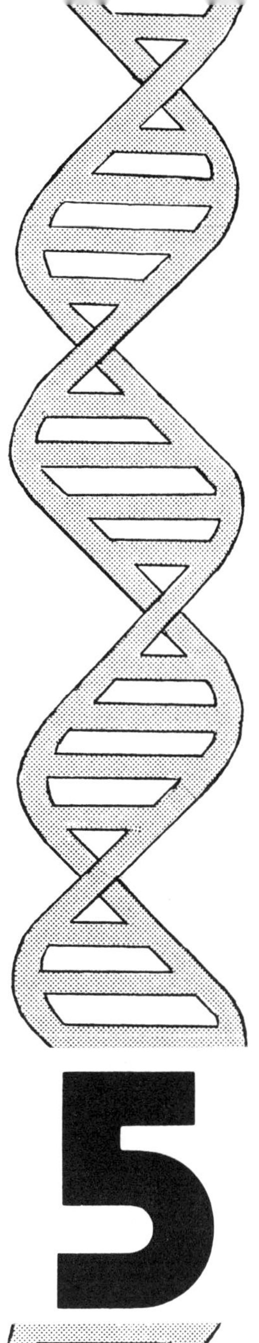

5

Recombining Genes for Medicine

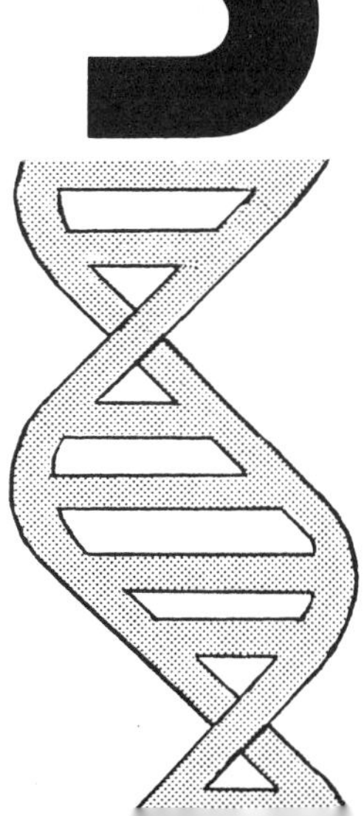

The genetic assault on disease moves forward on many fronts. Interferon, despite some setbacks, is still being considered as a possible anticancer drug. Interferon is produced naturally in the human body and is believed to be the body's first line of defense against viral infections. It was discovered about twenty-five years ago by the British virologist Alick Isaacs and his Swiss collaborator Jean Lindenmann at the National Institute for Medical Research in Britain. Since people contract only one viral disease at a time, it appeared that the presence of one virus interferes with the ability of others to grow and cause disease. No one knew the reason for this, but after a year of searching for an answer to this problem, Isaacs and Lindenmann discovered that a chemical manufactured by cells that have been attacked by viruses halts further attacks by other viruses. They named this chemical that had the ability to interfere with virus infections "interferon."

Some researchers struggled at that time and through the next ten years to produce small amounts of interferon. In the late 1960s, Finnish scientist Kari Cantwell introduced a better method of producing this much-wanted research chemical, but even with this improved method, the substance that was produced was only 0.1 percent pure. Technicians had to work for many months to harvest 100 million units of interferon from blood cells of humans. A treatment course for various types of leukemia might require a dosage of 10 million to 30 million units twice a week for several months. Since interferon is specific for species, interferon from laboratory animals could not be used to add to the supply needed for clinical trials with humans. However, some earlier successful experiments with the effect of interferon on mice tumors encouraged the scientists to persevere. By the early 1970s, enough interferon was produced to carry out a few clinical experiments.

Genetic engineering enabled scientists to produce more than several hundred billion units of virtually pure interferon in one week, and methods of production could be scaled up as required. A number of established pharmaceutical firms worked closely with biotechnical companies such as Biogen in Switzerland, and Genentech and Cetus in California to clone interferon.

Interest in interferon is based on its possible value in the control or cure of a number of diseases. While scientists have had antibiotics that could attack many bacterial diseases since the early 1950s, they had no such "magic bullets" to attack most diseases that are caused by viruses. Interferon has proved effective in the treatment of viral diseases such as chicken pox, shingles, the common cold, and some others.

One of the most exciting avenues of research with interferon is its use in the war against cancer. Numerous cancer patients have been treated either with the interferon produced by conventional methods such as that of Cantwell or by interferon produced through the use of cloning. Some of the results are encouraging, but others are disappointing. Although it was hoped that the pure variety would not produce side effects, such has not been the case. However, since cancer is not just one disease, but a term used for a collection of about one hundred different diseases with a similar manifestation, it is not surprising to find that interferon is not a "magic bullet" that destroys all tumors and other forms of cancer. The work continues, but many scientists think it is unlikely that interferon by itself will be an outstanding cancer cure. There is the possibility that through genetic engineering, specific forms of interferon may be tailored to destroy specific tumors. Even though interferon is no longer considered the miraculous cure some hoped it to be, it is still being researched as an anticancer and an antiviral agent.

In the summer of 1982, British scientists reported they had evidence that a nose spray containing interferon could prevent at least one type of the common cold. Earlier studies had not used interferon produced by genetic engineering, and it was not certain that the kind produced by genetically altered bacteria would be effective. At the Medical Research Council Common Cold Unit it Salisbury, England, nineteen volunteers using the interferon nasal spray remained cold-free after exposure to the kind of virus that is responsible for about 25 percent of the colds in adults. Eight of twenty-two volunteers who were given a nasal spray that was labeled as containing interferon but actually did not, developed colds after they were exposed to the same virus. There were no observable side effects from the nasal spray containing interferon. Scientists are optimistic that the spray will work well on a variety of cold-causing viruses. Eventually, it might be used by surgery patients, the

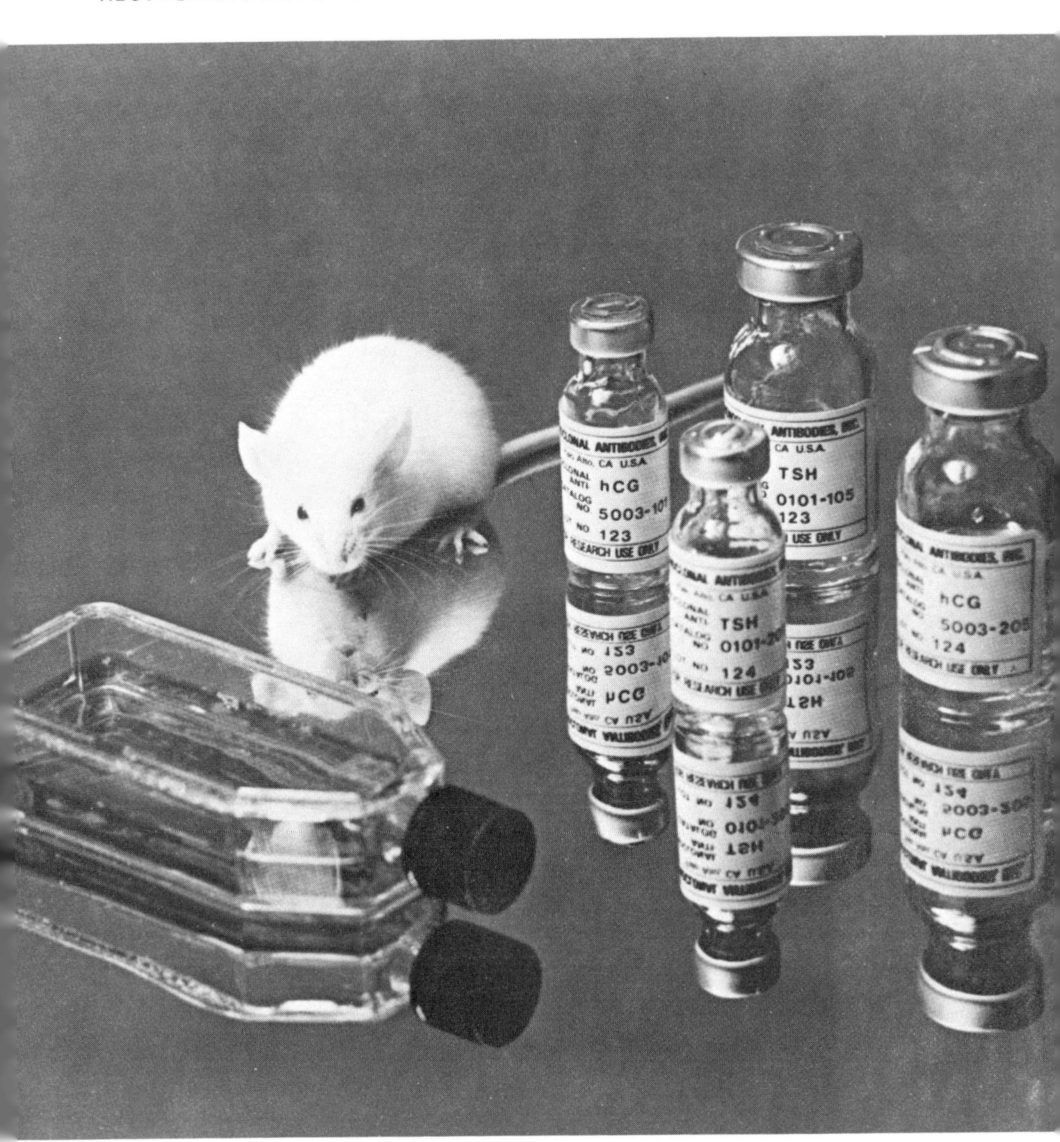

Figure 5-1. Cells from mice and people help to develop drugs that fight disease. *Monoclonal Antibodies, Inc.*

elderly, and others who might have serious health problems it they were to contract colds.

Genetic engineers had discovered about twenty different interferons by the end of 1981. Interferon is now known to be involved in functions other than the body's network of defense against disease. It has been helpful in decreasing the severity of chicken pox in children who are suffering with leukemia. This is more important than it may seem, since these children's immune systems have usually been weakened by chemotherapy, and they can become extremely ill or even die from chicken pox. Interferon may possibly be involved in nerve function, growth regulation, and the development of the fetus. Some studies indicate that it may be helpful in slowing the progress of multiple sclerosis.

The increased supply of interferon through cloning is giving medical researchers the opportunity to experiment with the drug in many areas. They have just begun to unravel its secrets. Doctors do not know much about the ideal dosage or the length of time to continue treatment. Although there have been some disappointments, cloned interferon may well play an important role in medicine of the future.

In 1981, DNA cloning was reported to produce a limited amount of vaccine that is effective against a disease known as hepatitis B, or serum hepatitis. This work that began with research at the University of California at San Francisco may have far-reaching importance. Hepatitis B is a major world health problem that affects about half a billion people and is carried by about 200 million people worldwide. Estimates of the number of carriers in the United States range from 700,000 to 1 million.

Hepatitis B is a viral infection of the liver that destroys liver cells. People with this disease are at risk of developing liver cancer. Symptoms of hepatitis range from the most mild, which often escape detection, to the most severe, where the disease is fatal. In general, after people have been infected, most of them regain their health and acquire immunity, but about one in ten becomes a chronic carrier because the hepatitis virus remains in the liver cells. Prevention by vaccine may eventually be one contribution to the control of hepatitis B. The detection of carriers is already possible through a complicated technique involving radioactive viral DNA.

One interesting aspect in work with the hepatitis gene is the use of yeast instead of bacteria in cloning the desired gene. The yeast adds sugars to proteins and thus produces complex biochemical molecules that are valuable in the process. Another benefit of using yeast is the easy adaptation of other methods for making the gene on a large scale. Techniques for growing large amounts of yeast have long been developed commercially to make beer, wine, and bread. And since the yeast does not produce certain toxic chemicals that are a problem in the fermentation of bacteria, it is of special interest to pharmaceutical firms that are developing new drugs by cloning.

Sometimes genetic research that seems unrelated to medicine leads to a greater understanding of human disease. For example, Dr. Barbara McClintock's work in corn fields and laboratories over a period of many years led to the discovery that not all genes are held firmly in place in chromosomes like pearls on a necklace. The mobility of genes may be involved in the transformation of normal cells into those that form cancers, in virus-transmitted diseases, and in resistant bacteria. Although the importance of her work was not recognized for many years, Dr. McClintock was awarded the Nobel Prize in December of 1983.

Figure 5-2. A biochemist working with an autoradiograph of RNA molecules from yeast cells.
David Powers, UCSF

The whole problem of infectious diseases presents a tremendous challenge to the scientists who work with genetic engineering. Such great strides have been made in curing infectious diseases that the need for more medicines to cure them and vaccines to prevent them is often underestimated. More than 500 infectious diseases are recognized in medical textbooks, and as many as 200 of these cannot be cured or prevented. Infections are among the most serious complications in the treatment of cancer and other diseases because the immune system that is crippled by drugs used in their treatment cannot fight infections that seem minor in healthy people. The problem of chicken pox in children with leukemia is a good example of this, as mentioned earlier.

Even those diseases that have long been curable with antibiotics may again become a threat if research does not continue to hold infectious disease treatment as a priority. For example, some diseases that yielded to penicillin when it was first introduced to patients at the time of World War II are now resistant to it unless doses are increased by a factor of 40.

Infections may present a serious threat to health in still a different way. No one knows how many *non*infectious diseases actually involve microorganisms. Some scientists believe that they may play a part in the development of a wide variety of diseases such as arthritis, cancer, senile dementia, and diabetes. It is not surprising to find that there is great interest in the potential of gene splicing for the cure of disease.

Hormones are another aspect of the work in gene cloning for the field of medicine. In fact, the first drug to be developed through gene splicing and cloning that was approved by the United States Food and Drug Administration was insulin, a hormone used in the control of diabetes. There are an estimated 60 million people in the world who suffer from diabetes, but not all of them use insulin to keep their disease under control. Eli Lilly and Company, producers of Humulin, the genetic-engineered insulin, estimate that about 35 million diabetics live in underdeveloped countries where they are neither diagnosed nor treated. Of the 25 million who live in developed countries, only about 5 million are treated with insulin. Some cases

Figure 5-3. A worker inspects fermentation tanks containing nutrient solution and *E. coli* bacteria which produce Humulin. *Eli Lilly and Company*

of diabetes can be controlled without the drug, but even in the developed world, it is believed that there are many cases of diabetes that are not diagnosed and many that are not treated.

Most of the insulin in use today is made from the pancreas glands of steers and pigs. It takes twenty steers and eighty hogs to harvest enough pure insulin for a single patient for a period of one year. Millions of animals are needed to meet the current need, and drug companies cite the fact that there may be a shortage of animal insulin in the future. The amount of meat consumption determines the number of steers and pigs available for the extraction of insulin, and diabetes is increasing in the human population at a faster rate than meat consumption. Diabetics who have the disease under control live longer and produce more children who transmit the defect than in the days when the disease was untreatable. All of this is important when one considers how much insulin is really needed and how much might be needed with improved medical care.

Beef insulin and pig insulin differ slightly in chemical composition from human insulin. Using a new chemical process, Novo Industri of Denmark has begun the manufacture of the equivalent of human insulin by modifying that from animal pancreases. This form of insulin was placed on the market in June of 1982 in Great Britain. This pure insulin now is available along with Humulin for those who suffer side effects from that derived directly from pigs and steers. These new products insure both a supply of insulin in the pure form and a supply that can be increased at a pace that will prevent any possible shortage in the future.

One substance that is being produced by genetic engineering is of special interest to families in which a child does not grow to nearly normal height. There are as many as 15,000 children in the United States who suffer from a condition known as hypopituitarism. No one knows why their pituitary glands do not produce the normal amount of hormone necessary for the growth of bones and other tissue. Before the days of genetic engineering, human growth hormone was extracted from the pituitary glands of cadavers, and it took fifty glands to yield enough hormone to treat one child for one year. Treatments cost about $10,000 a year for one patient, and were usually needed for a period of ten years. There was not enough hormone available for all who needed it, even if their families were able to afford the treatment. Many children who could have reached normal height remained dwarfed.

Today, human growth hormone is being made by recombining DNA and cloning. In the laboratories of Genentech, Inc., the gene that orders the production of growth hormone is inserted into the DNA of *E. coli.* When this bacterium reproduces, the gene for human growth hormone is transmitted to the millions of offspring.

This is the basic outline of the way human growth hormone is produced at Genentech: The process begins with a single *E. coli* cell. DNA coded for the production of human growth hormone is inserted into a plasmid taken from the *E. coli* cell. The plasmid is then put back into the cell, which is grown, using standard microbiological techniques, until there are several milliliters of fermentation broth. At this point, there are about 1 billion *E. coli* cells in the broth. This is added to a sterile fermentation medium and grown for five to ten generations. (One generation lasts only forty minutes.) Each bacterial cell carries the human growth hormone inside it, so there is now a considerable amount of the desired material.

The entire fermentation process is carried out in a totally contained stainless steel fermentor. Appropriate precautions are taken, such as sterilizing exhaust air, to keep the culture pure. Next, the recombinant culture is killed in the fermentation apparatus, and it is harvested using a centrifuge. The dead *E. coli* cells containing growth hormone are broken open and further purified. The final product is then checked for purity, and it is identified to make certain that the desired product is present. All through the process, there is careful control of what is happening, records are kept, and the entire process can be exactly reproduced.

In a laboratory where work is being done with genetically engineered products such as human growth hormone, the technicians do not wear the usual white lab coats. Their entire bodies are covered with sterile white gowns, head and face gear, footcovers, glasses, and gloves. Every niche of the manufacturing area is immaculately clean so that no foreign matter can get into the product.

Clinical trials with human growth hormone have already shown exciting results. For example, in one case a twin stopped growing while her sister continued to grow at a normal rate. Injections helped the shorter girl to catch up. One ten-year-old boy, who looked like a six-year-old because of his lack of height, grew three quarters of an inch between October and the following January when given the growth hormone. Many children who are shorter than 97 percent of the children in their age groups have begun taking this hormone and are growing at a much faster rate than in the months before they began the treatment.

Some children are extremely short because of lack of growth hormone; others stop growing for different reasons. Malnutrition is a common one. Since injections of growth hormone appear to have a desired effect even in those children who seem to produce normal amounts of growth hormone, might some parents try to use it just to increase the size of their normal children? Researchers stress that incorrect dosages could lead to complications such as diabetes and cardiac problems. Since the injections must be administered three times a week over a period of five to ten years, they suggest it is unlikely that anyone other than significantly short children will want the treatment.

Researchers hope that human growth hormone may be useful in treating a wide variety of conditions. For example, it may be helpful to older people who suffer from bone decalcification, in treating bleeding ulcers that do not respond to other treatment, and in promoting healing of wounds, burns, and broken bones.

The use of recombinant DNA developed by genetic engineers is just beginning to reach the practical side of medicine. The world's first pharmaceutical product of recombinant DNA, Humulin, reached the market in 1982. By the end of the next year, a new process using recombinant RNA was developed that may allow scientists to produce insulin more cheaply, in purer form and greater quantities than through the use of recombinant DNA.

The potential of new medicines such as Interleuken-2 (IL-2) excites genetic engineering companies. IL-2 is normally produced by the body in small amounts as part of the immune system. Through the use of cloning techniques such drugs can be made available in sufficient quantities for experiments in the treatment of cancer and other diseases that may be caused by a breakdown in the body's immune system.

A recent breakthrough in understanding how the immune system works is the isolation of a protein that regulates the body's response to tissue transplants, infectious agents, and tumor cells. The protein is the specialized receptor molecule on T-cells, one of the two major cell types that control the body's immune response. The receptor allows the cells to recognize foreign substances. This new knowledge will help develop therapy for immune system disorders and for autoimmune diseases, better vaccines for infectious diseases, and more successful tissue transplants.

With such promising developments, what can one expect in the next ten or twenty years?

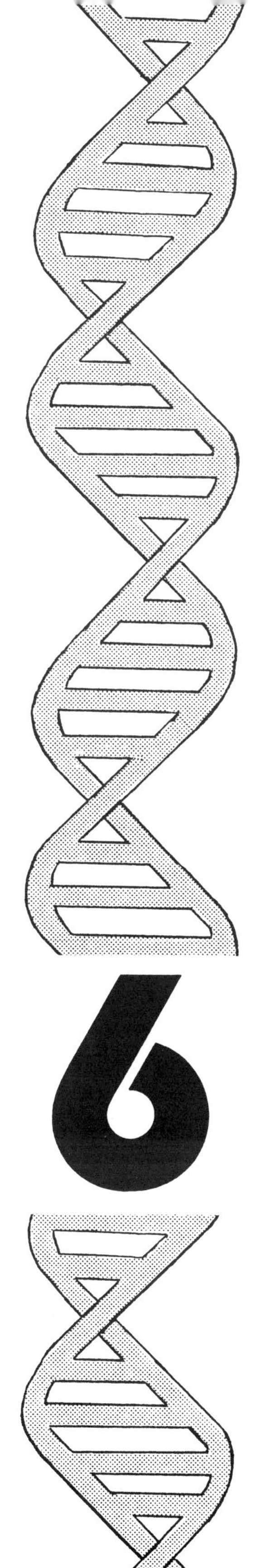

6

Monoclonals: Super Antibodies

Figure 6-1. A Greek mythological chimera was made up of parts of a lion, goat, and serpent. Today, a chimera may be formed in the laboratory by fusing a mouse cell and a human cell.

The fire-breathing chimera of Greek mythologists may have heralded one of the most important tools of modern medicine. This imaginary animal composed of various parts of a lion, goat, and serpent was far different, however, from the chimeras that are being created in the laboratories of today.

One modern version of a chimera is a cell which combines a mouse cell and a human cell in such a way that the new cell produces a specific antibody. The cells are then cloned, producing a single line of pure antibodies. These monoclonal antibodies, as they are called, already are used in diagnosing cancer and other diseases, in treating some viral diseases, and in purifying some rare and valuable chemicals that are used in biology.

Antibodies are natural substances that are secreted by your body's plasma cells in response to an invasion of something that your immune system recognizes as foreign. They are the first line of defense against invading organisms, foreign tissues, and toxic chemicals.

Antibodies recognize the shape of chemical markers on a foreign substance, which is known as an antigen. For example, most bacteria wear a membrane coat studded with proteins that act as tags. Cells in the human immune system use these tags to identify the foreign bacteria. Antibodies combine with the bacteria, or other antigens, in a lock-and-key fashion, setting in motion processes that can neutralize and eliminate the foreign substance. So the natural function of an antibody might be thought of as clamping onto an intruder, such as a virus, bacterium, transplanted organ tissue, or foreign protein, and making it inactive.

The antibodies produced by the body usually win in their battle against the antigens, but sometimes they do not. There may be too many cancer cells, or too much of a virus or of any antigen. Since injections of antigens into an animal cause the formation of antibodies, vaccines can be developed from these antibodies for use in humans. Injections of some kinds of killed viruses or purified viral products protect you from invasions of live viruses of the same kind. This is the principle behind the use of vaccines.

For example, if you are vaccinated for measles and the measle virus multiplies in your body at a later date, you do not get sick because your immune system has been prepared to act against this infectious agent. But it is impossible to use vaccinations for all diseases. Not all viruses can be isolated and grown in the laboratory, and there is always concern that some live virus might sneak through the killing process and cause the disease it is meant to prevent. And considering the number of different kinds of diseases, it is obvious that one cannot be vaccinated against all of them.

Scientists have long dreamed of using antibodies in specific ways to help fight disease when it strikes, and they have done so to a very limited degree using older methods. Before the days of monoclonal antibodies, scientists would inject an antigen, such as human lung cancer cells, into a laboratory animal and wait until the animal made antibodies against it. Then they would separate the antibodies from the animal's blood and use them in their patients. However, a huge variety of antibodies is produced whenever an antigen is introduced into an animal. So many different kinds of antibodies were produced in the test animals that it was virtually impossible to isolate the specific one that was desired.

In addition to the problem of the hodgepodge of antibodies, antibodies differ from animal to animal and from time to time in the same animal. So isolating a particular antibody to fight a certain disease seemed a hopeless dream.

In another approach, scientists were able to grow antibody-producing spleen cells from immunized mice, and many of these cells would manufacture specific antibodies. However, this approach was not very successful because the antibodies that were produced only survived a few weeks and the amount that was produced was very minute.

So there were two problems that had to be overcome before one could hope to fight disease with antibodies that were produced outside a patient's body: Antibodies that targeted a specific antigen were needed, and the antibodies had to be long-lived if they were to be useful. The answer came with the development of monoclonal antibodies.

The first monoclonals were made in 1975 by César Milstein and George Köhler at the Medical Research Council in Cambridge, England. These scientists succeeded in fusing spleen cells from a mouse that had been immunized with a particular antigen to cells from another mouse that had a form of cancer known as myeloma (a cancer of the blood-forming tissue). The result was a hybrid cell that produced a single pure antibody and was immortal because it came from a tumor cell. Such a fused cell is called a hybridoma. In other words, a hybridoma is a hybrid cell that contains the genetic machinery for making an antibody and the cancer genes that make the cell immortal. The hybridoma will continue to divide forever, making clones of itself that will produce a single pure type of antibody.

Today, the laboratory technique for making hybridomas is a well-established procedure. After the tumor cells from a patient have been injected into a mouse and the mouse's spleen has produced antibodies against the cancerous cells, scientists remove the spleen and tease the cells apart.

In sophisticated laboratory procedures that are carried out under sterile conditions to prevent bacterial contamination, the cells containing antibodies are mixed with myeloma cells plus the chemical used for car antifreeze, polyethylene glycol (PEG). Together they are spun in a centrifuge. The PEG chemically dissolves the outer membranes of some of the cells in such a way that the cells stick together. After these cells are washed, a small percentage of them remain stuck together and they begin to fuse. Only about 1 in 20,000 cells fuses successfully. All the cells are put on a growth medium that permits only the fused cells to grow. After a period of several weeks, clones of the fused cells are checked to see if they are producing the desired antibody. Those hybrid cells which do are used as antibody factories.

One of the most exciting potentials for monoclonal antibodies is using them as "bullets" to seek out cancer cells. If antibodies can be prepared to attack a specific kind of cancer from which a patient suffers, they can be injected into the bloodstream to pursue the cells of spreading tumors. Some scientists feel that the greatest use of monoclonal antibodies will be to clean up cancer cells that have been missed by surgery, chemotherapy, or other treatments that are already being used to get rid of the major part of the cancer. If you know a person who is recovering from cancer surgery, you are probably familiar with the question, "Did they get it all?" Monoclonal antibodies may someday be used routinely to destroy any cancer cells that may remain in a patient's body, so there will be no chance that stray cells might initiate cancer in the same place or in a different place.

Since monoclonal antibodies have the ability to seek out and attack cancer cells without harming normal cells, researchers hope that they can find a way to put them to use in treating a wide variety of cancers. Many doctors believe that someday there will be "magic bullets" made of antibodies that are perfectly matched to different kinds of cancer. This would eliminate the need to destroy large numbers of normal cells in order to destroy cancer cells.

In one of the first successful uses of monoclonal antibodies in humans, doctors in California treated a sixty-seven-year-old

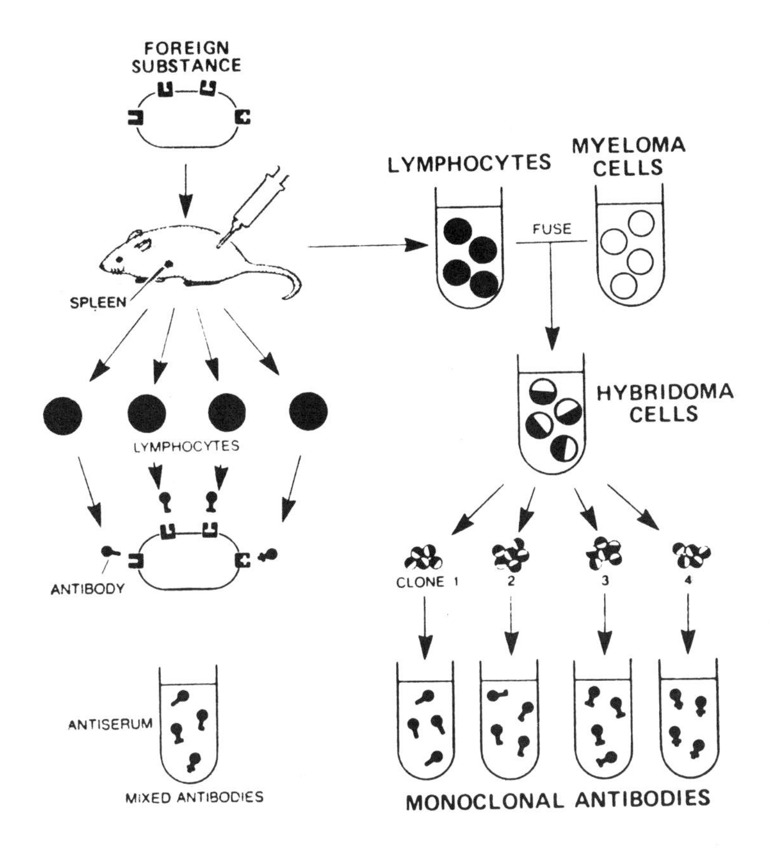

Figure 6-2. A diagram of how monoclonal antibodies are produced by cloning hybridomas. *Stanford Medical Center News Bureau*

man who was suffering from an unusual form of leukemia-like cancer known as lymphocytic lymphoma. He had been treated for two years with standard anticancer drugs and with experimental doses of interferon, but the cancer had not been brought under control. Since his cancer was producing a unique protein, scientists at Stanford University were able to make a monoclonal antibody that was effective against his particular type of malignant lymphocytes. This took six months of painstaking research.. The monoclonal antibodies were injected into his bloodstream in a series of eight doses in which the amounts were carefully increased. By the third and fourth injections, his fever and night sweats (symptoms characteristic of lymphoma) had disappeared. His lymph nodes gradually became smaller, his liver and spleen returned to normal size, and the tumors on his scalp disappeared. The patient remained in remission—without symptoms of the disease—with no further treatment.

Although this case appears to illustrate successful treatment in one patient for at least a year and a half, doctors say that it takes many cases and periods of five to ten years to determine the true success of a technique. Actually, research is moving forward in this area, but there is no certainty that what helped one patient may help others. And any treatment that could be used for a large number of patients must depend on a simpler technique than that for the man in the case described above. Many scientists doubt if the laborious process to obtain an effective monoclonal antibody can be streamlined.

Another exciting but limited approach to treating cancer with the use of monoclonal antibodies is in the area of bone marrow transplants. This approach is being tested at Harvard Medical School and Dana-Farber Cancer Institute. Blood cells are manufactured in the bone marrow, so if this tissue becomes cancerous, a person suffers from leukemia. Doctors are using monoclonal antibodies to help leukemia patients who have no bone marrow of matching tissue type available to them for transplants. In this technique, a patient may be brought into remission by massive doses of radiation and drugs. Then some of the patient's bone marrow is removed, treated with monoclonal antibodies to destroy every trace of cancer cells, and frozen so that it can be used when the patient's disease becomes active again.

Monoclonal antibodies appear to provide an effective, practical treatment for the rare but serious disease known as SCID (severe combined immunodeficiency). In 1982, it was reported that a baby girl who was born without resistance to disease was successfully treated through techniques that were developed by genetic engineers. Doctors at the Dana-Farber Cancer Institute believed they had cured the baby, who had been required to live in a bubble because her body lacked the natural immune elements needed to resist disease. Efforts to treat such patients with bone marrow from donors usually fail because of difficulty in matching tissues. Immune cells of the donor can attack the patient's cells as foreign objects. But with monoclonal antibodies, this problem can be avoided.

In cases such as the above, early results have encouraged researchers to continue their experiments in the hope that laboratory-made monoclonal antibodies will provide some important help in the attack against cancer. Years of research and large amounts of funding will be needed for further experimental studies before the full potential of monoclonal antibodies can be translated into numerous clinical trials. Procedures must be widely tested before they are used in widespread treatment. But monoclonal antibodies are being hailed as exciting new tools in a wide variety of areas.

For example, scientists have succeeded in experiments that made hybridomas using two kinds of human cells rather than a mouse cell and a human cell, producing a human-human hybridoma. Dr. Robert Lundak, an immunologist at the University of California at Riverside, is experimenting with antibodies that could be used in experimental treatment of jaundice, a blood disease that causes a yellowish skin in newborn babies, and of childhood leukemia. Although many problems remain, such experiments provide hope for the future.

A number of companies are already producing monoclonal antibodies commercially. For example, Monoclonal Antibodies, Inc. devotes part of its efforts to research and development. This includes the development of new hybridoma cell lines and the maintenance of existing cell lines, the design and development of new diagnostic kits, and the production of selected monoclonal antibodies for research purposes.

Vega Biotechnologies, Inc. is another of the companies that works in the field of monoclonals. One of their products, a diagnostic kit, is used in clinical laboratories to accurately measure small amounts of various substances in blood samples. Using antibodies, the tests can measure biological substances in amounts as minute as one trillionth of a gram. Knowledge of certain factors in the blood aid in the diagnosis and treatment of conditions such as diabetes mellitus, cirrhosis, hepatitis, and thrombosis.

Monoclonal antibodies may someday be valuable in detecting cancers and other diseases in high-risk individuals before there are any clinical signs of illness. They may help to make blood typing more accurate, and in the typing of body tissues for a better match between donor and recipient.

One of the most exciting areas of monoclonal antibody research is in the field of controlling diseases of the immune system. Although this system makes it possible for people to live in a world that is highly populated with disease-producing organisms, sometimes the system goes awry and attacks organs and tissues in a person's own body. Such errors seem to be involved in diseases such as allergy and arthritis. Monoclonal antibodies that could seek out the faulty cells of an immune system could stop these cells from attacking their own tissues.

These are just a few ways in which monoclonal antibodies show promise. They may even be helpful in conjunction with new recombinant DNA techniques. In one step, scientists have already made interferon 5000 times more pure by mixing the drug with its antibody. No wonder the president of Hybridtech, Inc., one of the firms producing monoclonal antibodies for diagnosis and research, believes that we are sitting on the brink of a medical revolution.

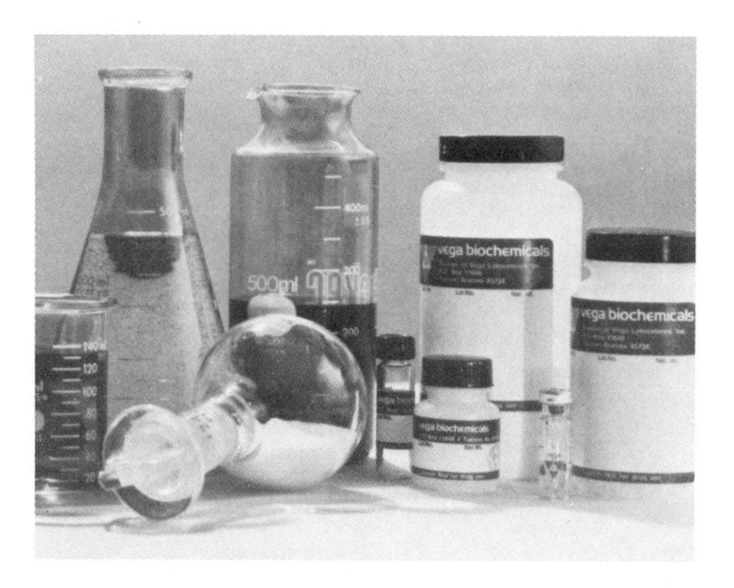

Figure 6-3. Many monoclonal antibodies are produced commercially.
Vega Biotechnologies, Inc.

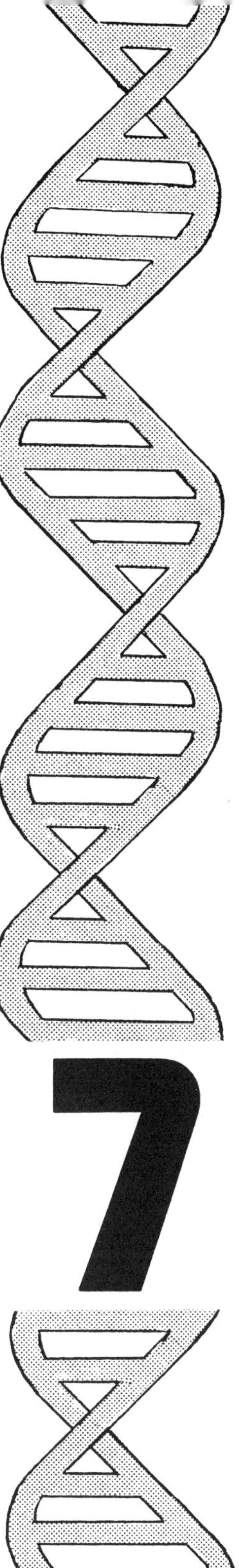

7

The New Genetics and the Farmer

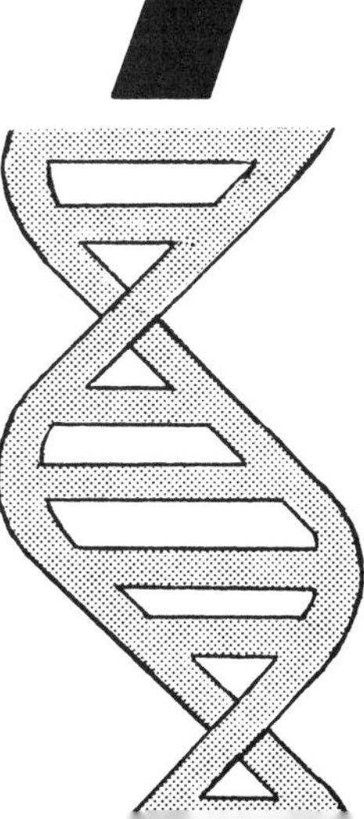

Although the number of people that can be fed by the American farmer continues to increase, the need for more food also continues to grow. The increase in the amount of food produced per acre is slowing, and the cost of water and fertilizer is rising. In the United States alone, three million acres of farmland are lost each year through a combination of urban sprawl and soil erosion. New and exotic insects nibble at the crops. Problems of water supply and salt in the soil make it urgent that agricultural research be given high priority.

Although farmers in distant places were helped by the Green Revolution of the 1960s, so called because newly developed hardy, high-yielding strains of wheat and rice were made available, few farmers in the developing countries can afford the chemicals and machinery necessary to grow these crops. One fifth of the world's population is estimated to be suffering from malnutrition. Consider how the problem can increase if the population grows from four billion to seven billion people by the year 2000. By then, one third of the world's arable land will probably have been destroyed. If world food supplies are not doubled by that time, people in the less-developed countries are expected to suffer from starvation on a massive scale.

Geneticist Ronald Phillips of the University of Minnesota believes that all of the food produced in the last 1100 years would be enough for the needs of only the next 40 years. He and other geneticists are among those working toward meeting the challenge.

New strains of plants come from new combinations of genes. Bees and other insects have been exchanging genes between plants for millions of years by carrying pollen on their bodies. People breed plants by hand, exchanging genes between closely related plants. But without the techniques of the new genetics, this transfer of genes is limited to such plants.

Many people feel that there are breathtaking possibilities for farmers of the future because of new research in manipulating the genes of plant cells and growing these cells under sterile conditions. There may come a day when plant breeders can order certain genetic traits from the laboratories of the "gene splicers." Scientists are also developing technology for growing plants from cell cultures in large quantities so that they can be harvested commercially. However, progress in the genetic engineering of plants has been slow; it may be many years before it supplies large-scale help to the farmer.

Although experiments are still in the early stages, scientists are working with genes to increase the disease resistance and stress tolerance of plants, and to improve the yield and the nutritional quality of major crop plants. However, by various methods, plants have developed natural resistance to certain diseases, and apparently these diseases have been wiped out. This resistance is transmitted genetically. When altering a plant's genetic makeup, care must be taken not to remove resistance to these diseases, which may one day reappear.

Consider the importance of the development of new strains of plants that can thrive in salty soils or under salt-water irrigation. About 500 million out of the 3500 million acres of the world's arable land are irrigated, and that land regularly accumulates salt from evaporation.

The head of barley that is seen in the left side of the picture on page 79 is from a plant irrigated entirely with sea water, while the head of wheat on the right is from a plant irrigated with fresh water. The barley would have grown stronger with fresh-water irrigation, but in the future, new strains may grow well in salty water or may be able to live for weeks without any water.

At the University of California at Davis, scientists have succeeded in cloning a microbe with a gene that tolerates sea water. Another team of scientists at the same university has developed a cherry-size tomato that grows in sea water. Eventually, it may be possible to use recombinant DNA technology to transfer the gene or genes responsible for salt tolerance in marsh grass to salt-sensitive crops like wheat. The use of gene splicing to produce such new strains of plants may help to increase the food supply.

Many plant scientists are counting on the unusual bacterium *Agrobacterium tumefaciens* to piggyback, or carry, genes from one unrelated plant to another. This bacterium causes certain plants, such as tomatoes, to grow tumors, called galls, by adding a piece of its own DNA to plant chromosomes. It infects about 10,000 different kinds of plants. By inserting new genes into the bacterium's DNA, scientists hope to transfer genes from many sources into plants to improve them in a wide variety of ways.

Figure 7-1. The head of barley on the left is from a plant irrigated with sea water. The barley on the right was irrigated with fresh water.

International Plant Research Institute

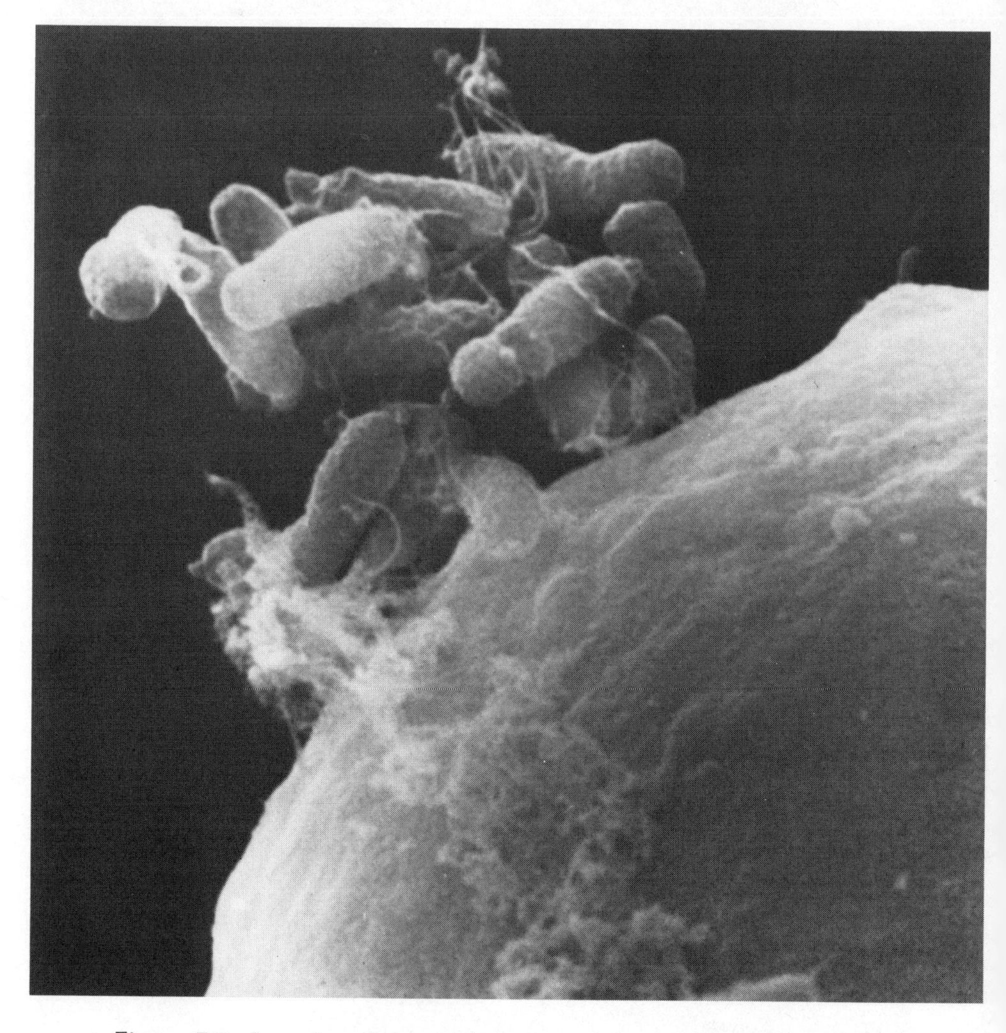

Figure 7-2. Scanning electon microscopy shows several *Agrobacterium tumefaciens* as they begin to infect a carrot cell. In the process, the bacteria's genetic material will enter the plant cell. *USDA Agricultural Research Service*

A group of University of Wisconsin scientists led by John D. Kemp and Timothy C. Hall, formerly of the USDA Agricultural Research Service, has worked with the transfer of a gene from the French bean into a sunflower cell to create a "sunbean." The gene transfer was accomplished through the use of a ring of genetic material, a plasmid, found in *Agrobacterium tumefaciens*. The sunbean was introduced in June of 1981 in the form of plant tissue, not as a full-grown plant. However, it was hailed as an important first step.

In January of 1983, another landmark was passed. Scientists in Europe and the United States announced simultaneously that they had succeeded in getting a gene into a plant cell and having it expressed. Scientists used *Agrobacterium tumefaciens* to carry the gene that they wanted to transfer into plants such as petunias, sunflowers, tobacco, and carrots. They silenced the tumor-causing genes of the bacterium but kept the "carrier" genes that allowed the tumor material to be inserted into the plant without being rejected. The exciting part of these experiments was the fact that the transplanted genes were taken up by the plants into which they were transferred and they continued to operate. Hopefully, such work will lead to the day when farmers can grow crops of normal plants with foreign genes that protect against disease or that have other special desirable characteristics. Scientists working at the International Plant Research Institute predict that some improved plant varieties may be on the market by the mid-1980s.

One of the most important goals of genetic engineers is transferring the genes for nitrogen-fixation from legumes, such as peas and beans, into cereal plants so that they, like the legumes, can make their own fertilizer. Such an accomplishment would be a major breakthrough in food production, another Green Revolution. Recently, a bacterium found in China has played an important part in helping researchers at the USDA Agricultural Research Service to understand the genetics of the nitrogen-fixing *Rhizobium* bacteria. When *Rhizobium* bacteria invade the roots of soybeans, the roots swell, forming nodules. The soybean plant provides nutrients for the growth of the bacteria, and the bacteria produce nitrogen-containing compounds which fertilize the plant. Experiments are generally hampered by the slow rate of growth of the bacteria native to the United States. However, some species of *Rhizobium* bacteria that were recently imported from China grow much faster than those found in the United States.

In the United States, soybean seeds are routinely coated with *Rhizobium* bacteria during planting to promote the formation of nodules on the roots. If genetic engineers can combine fast growth with effectiveness, companies could produce these bacteria for soybean farmers faster and more economically.

Genetic engineering has made corn one of the most efficient plants for converting solar energy into food, and new characteristics that may help corn breeders may continue to improve this crop. Corn accounts for about one quarter of the world grain production, and about half of the corn is grown in the United States. More than a million American farmers grow seven billion bushels of corn a year in cornlands that would cover about 70 million acres, an area the size of the state of Arizona.

No wonder scientists are excited about the newly discovered plasmids that may someday lead to more productive hybrids of corn and sorghum, another one of the world's major crops. When an Agricultural Research Service plant pathologist and a North Carolina State University geneticist explored the tiny energy-producing bodies called mitochondria in corn cells, they found plasmids that could be used to breed plants with desirable traits such as disease resistance, increased yield, and improved seed quality.

Figure 7-3. Scientists examine the stored genetic stock of the many parent lines and mutants used in the study of plasmids in corn.

USDA Agricultural Research Service

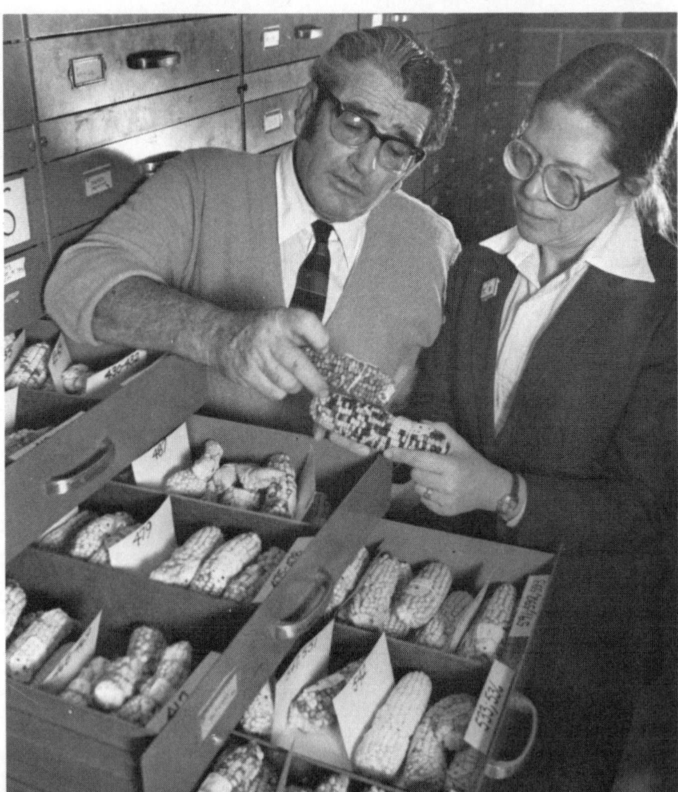

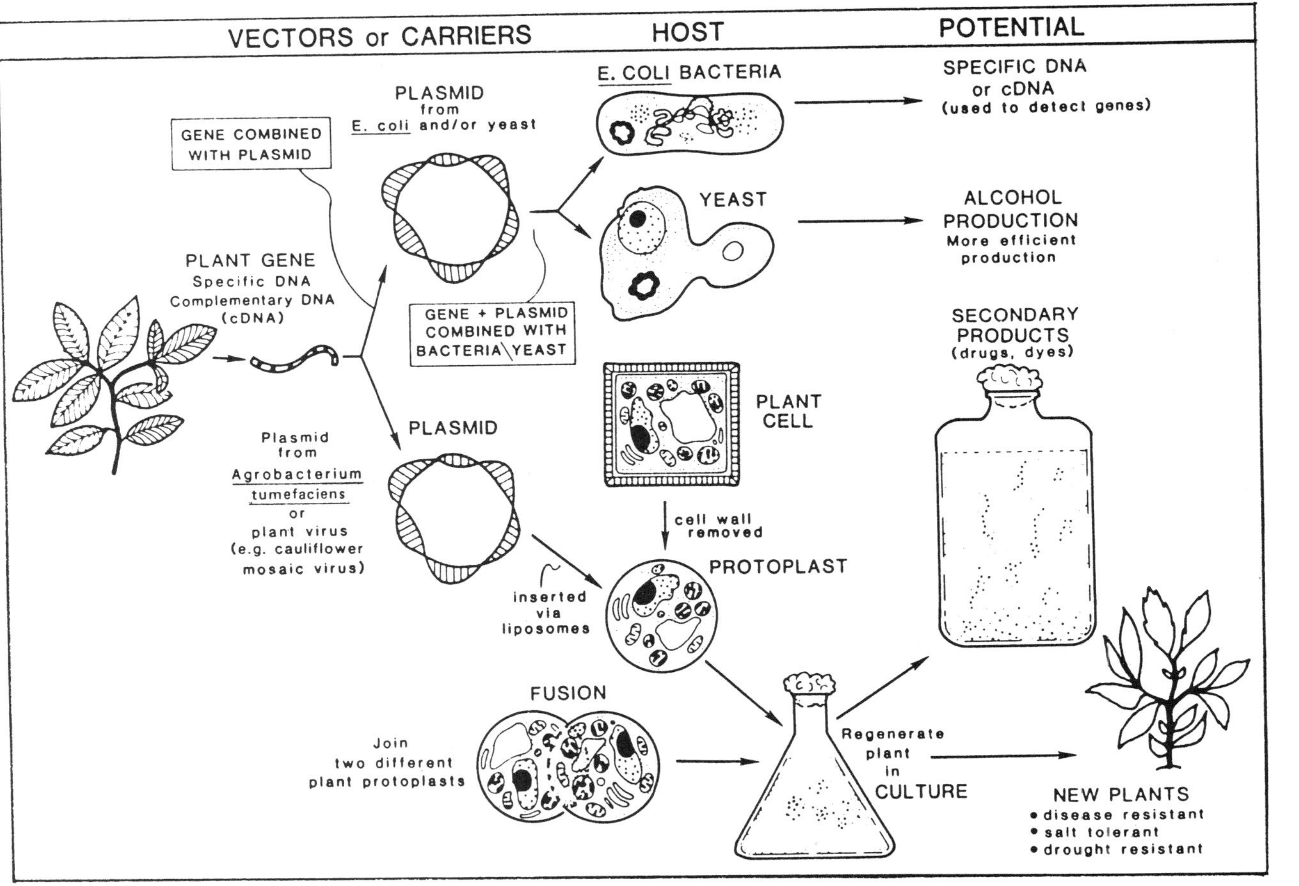

Figure 7-4. Some of the ways genetic engineering can improve plants. *USDA Agricultural Research Service*

Figure 7-5. Male sterile corn seedlings are prepared for genetic studies.
USDA Agricultural Research Studies

Scientists are working with plasmids to alter corn plants for genetically desirable traits such as male sterility. The tassel is the male pollen-producing part of corn. The female part is the ear of corn. Many breeders plant the types of corn they want to cross in adjacent rows, and then remove the tassels by hand from every second row. These detasseled plants are then fertilized by pollen falling from tassels in adjacent rows, thus producing hybrid seed. If male-sterile corn is planted, the laborious process of removing tassels by hand is eliminated.

The work with corn and many other plants selected for genetic engineering depends on the use of protoplasts in a tissue culture process. Plant protoplasts are individual plant cells whose walls have been dissolved by an enzyme. Through the use of chemicals, scientists have been successful in fusing some different plants that do not cross naturally. The full complement of genes from each plant are brought together and some of these combinations may be desirable. The selection of desirable plants can be carried on much faster using protoplasts in the laboratory than by growing plants in the field. And protoplasts may lend themselves to work with recombinant DNA since they can be manipulated more easily than other cells.

With proper nutrients and conditions, some protoplasts and other cells regenerate cell walls and divide to form clusters of cells called calli. These calli may develop roots and shoots and grow into mature plants. Geneticists are exploring the use of tissue culture with a number of different kinds of plants in hopes of greatly improving this promising approach.

Plants cloned in tissue culture are credited with helping to save the redwoods and in playing a large part in improving forests. Professor Ernest Ball of the University of California at Santa Cruz has been working with genetically identical "plantlets" grown from a single piece of redwood tissue for many years. In his experiments, the plantlets that have been successfully cultivated in a test plot are growing twice as fast as regular seedlings. These plants that began with tissue culture may live as long as 100 years before they are harvested. If they are preserved in a redwood forest, they may live for 800 or even 2000 years. In some experiments, 2500 tissue-cultured redwoods have been grown under field conditions so that they can be compared with those that were produced naturally.

Tissue culture breeding of genetically selected trees is important economically. Stabilizing traits and field trials can take fifty, or even hundreds, of years for many kinds of trees. Some loblolly pine and Douglas fir trees that are being cultured illustrate this. The number of trees that can be grown in 100 liters of media in three months is enough to reforest 120,000 acres of land at twelve- by twelve-foot spacing. No wonder scientists are excited about the cloning of trees in culture media.

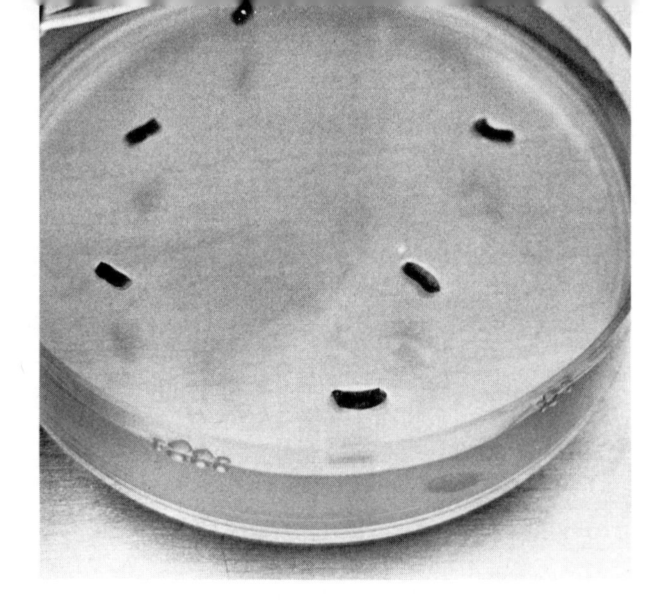

a

Figure 7-6. Tissue culture breeding of genetically selected trees begins with slices of tissue fragments set in a petri dish containing growth medium (a). After four to five weeks, the pieces of tissue will begin to differentiate into cell clusters, or calli (b). The developing green shoots are separated, then placed in individual vials containing root-inducing nutrients and hormones (c); these are later transplanted to forest acreage. Whole forests can grow from plants cloned in laboratories (d). *Weyerhaeuser Company*

b

c

d

Figure 7-7. Dr. Theodor O. Diener examines a potato plant for symptoms of potato spindle tuber disease, which is caused by a viroid. A new test for viroid detection has been developed using recombinant DNA technology.

USDA Agricultural Research Service

In 1971, U.S. Department of Agriculture plant pathologist Dr. Theodor O. Diener discovered a group of disease agents which he named viroids. Viroids are small naked molecules of RNA (ribonucleic acid). In the early eighties, he discovered a test for screening large numbers of potatoes for the viroid that causes potato spindle tuber disease. The disease is a threat to potato farmers in moderate climates and a very serious problem for those who live in tropical climates. When the viroid forces itself into the host potato, its genetic message becomes part of the plant in such a way that many more viroids are produced.

Viroids are unbelievably small, so small that it would take 30,000 viroids to weigh as much as the smallpox virus. But the potato spindle tuber virus can change potatoes from normal shapes to long ones with deep cracks and can devastate fields of seed potatoes. Other viroids damage a host of other crops. Few people doubt the value of genetic research with plants; many just wish it would progress at a faster pace.

Some of the applications of genetic engineering in the world of plants may sound simple, but they are the result of some extremely difficult and complex work. Desirable traits are often controlled by several genes located on different chromosomes. Even when a gene has been located and successfully transferred in a complicated procedure, the plant must be grown to full maturity if it is to benefit agriculture.

In animal genetics, one dramatic discovery that helps farmers has been the development of a vaccine to prevent foot-and-mouth disease in cattle and other animals. The new vaccine, reported in June of 1981, was the result of a joint research effort by the USDA's Agricultural Research Service and Genentech, Inc. Foot-and-mouth disease is a virus-caused illness of livestock that causes blisters in the mouth and nose and on the feet of cattle, sheep, goats, and swine. The excessive salivation due to blisters in the mouth may be the first sign of the disease. Strict screening and quarantine procedures have prevented major outbreaks in the United States since 1929. However, it causes major economic losses worldwide and makes diseased animals unacceptable for importation into disease-free countries. Although there were vaccines for this disease before the one produced by cloning recombinant DNA, they had to be refrigerated. This was difficult to do in developing countries. The new vaccine can be stored for long periods of time without refrigeration.

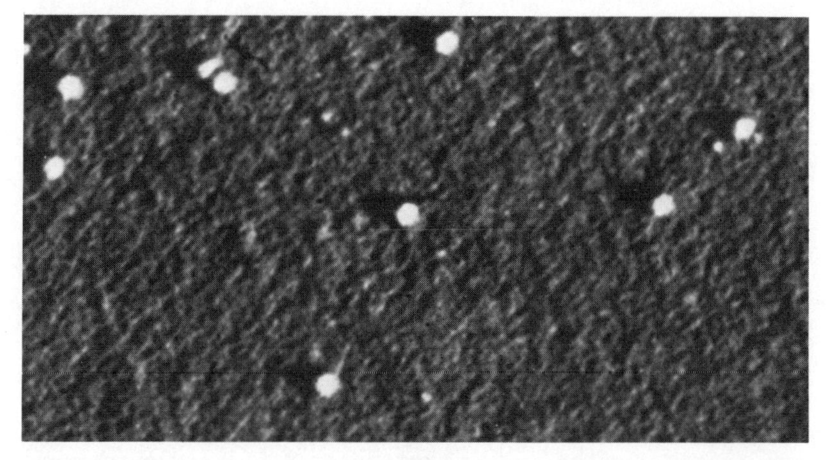

Figure 7-8. The foot-and-mouth disease virus is spherical in shape and about one-millionth inch in diameter. This organism is the smallest of any of the viruses affecting animals. *Courtesy of USDA*

Many scientists are hopeful that the revolution in animal genetics will enable farmers to increase the food supply dramatically. The sole function of nearly 80 percent of the cows in North America is to produce one calf a year. The conversion of feed into meat would be more efficient if the cattle could be made to produce twins or triplets through a cloning technique known as twinning. Some experiments to increase the number of young involve separating fertilized eggs at the two-cell stage or at a later stage when the embryo reaches the sixteen- to sixty-four-cell stage (morula). For example, eight pairs of identical twin mice have been produced by dividing morulae in half, and five pairs of identical sheep twins have been produced by dividing two-cell embryos in half. When these laboratory methods for producing clones by dividing early embryos become more practical, they may be used in experiments for the control of genetic variation.

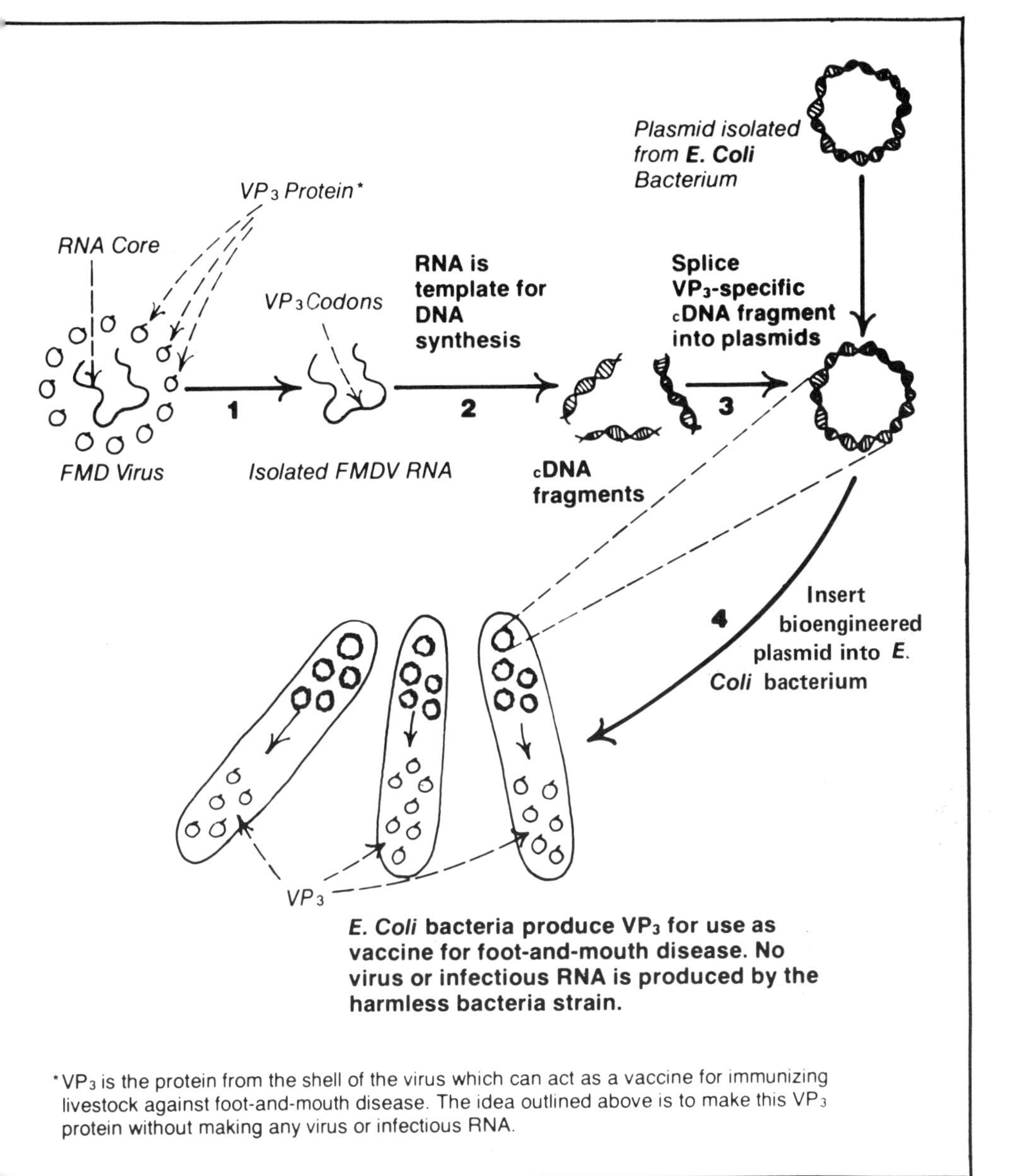

Figure 7-9. This chart shows how a vaccine for foot-and-mouth disease is made using recombinant DNA.

Courtesy of USDA

Embryo division and the growth of clones in surrogate mothers of closely related breeds may help to rescue some species headed for extinction. Estimates place the annual extinction rate as high as 1000 species per year. Experiments with embryo transfer are underway with some rare breeds, such as the Angora breed of sheep in Australia.

Although advances in the knowledge of gene transfer in farm animals are coming rapidly, work is still in the early stages. Foreign DNA is being injected into the fertilized eggs of experimental animals with the object of transmitting desired traits, but the genes do not always become part of a cell's DNA. When they do, it is only in a small percentage of the animals. However, scientists are optimistic about incorporating new genes into economically important animals so that supercows may produce more milk and fight disease more effectively. Someday, recombinant DNA may be important in producing more meat and better livestock products.

Estimates of the time necessary to develop cows, pigs, and sheep with new genes for desired traits range from two to twenty years. Autar K. Karihaloo, Director of Carnation Genetics, a division of Carnation Company, says that it boggles his mind to think what dairy cows will look like in twenty-five years.

Figure 7-10. These identical-twin calves are clones.

Colorado State University

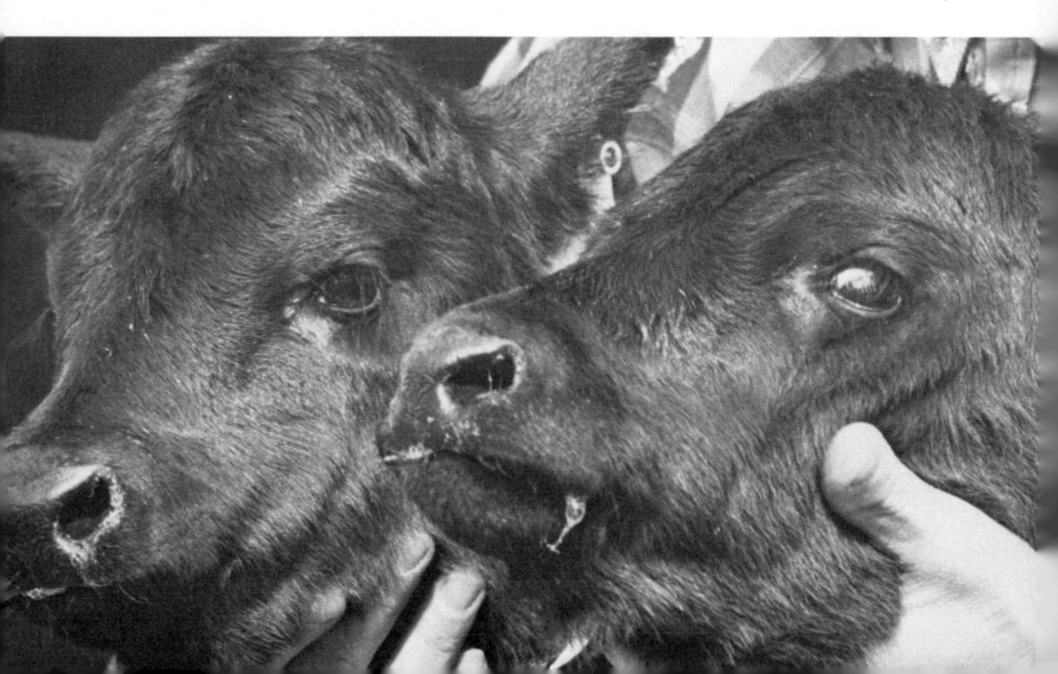

One of the outstanding researchers in the field of animal reproduction is George E. Seidel, Jr. who works at the Animal Reproduction Laboratory of Colorado State University. He points out that an experiment that at first seems totally useless may lead to important discoveries that help large numbers of people. For example, an experiment titled "Insertion of Wax Paper Between the Brain and Pituitary Gland of the Ferret" led to the discovery of a drug, gonadotropin releasing hormone, that is very important in the reproductive management of livestock. Many of the current projects in laboratories appear to have little value, but the accomplishments of Dr. Seidel and other researchers in producing genetically identical animals may lead to practical applications.

There are many more aids to agriculture on the horizon as a result of new techniques in genetic engineering. Commercial producers are investing in DNA technology in efforts to produce better vegetables such as supertomatoes and disease-free potatoes. Many ancient scourges of farming are being eliminated or will be in the future. Cloning and the new genetics holds great promise for the farmer.

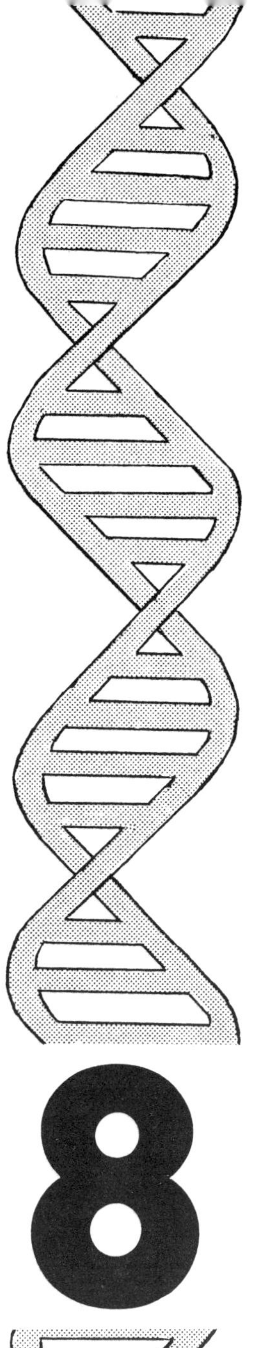

8

Cloning a Man?

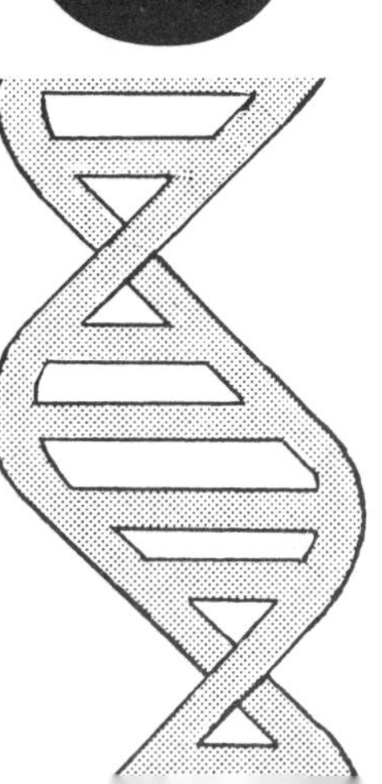

Those crazy scientists should not be playing God!

When they clone a man, I hope they choose a super-athlete.

I'd rather see them clone an Einstein.

They are cloning babies right now. How about that baby that began in a test tube? (This was not cloning. Many people confuse gene splicing and cloning with artificial insemination, in vitro fertilization, and other forms of reproduction.)

These are comments one can hear almost anywhere. They are made by people of all ages, and sometimes these people are well educated in areas other than genetics.

If you listened to comments about cloning a human that are made by geneticists you would probably hear some express boredom at the idea of cloning a man. Others would say the idea is ridiculous but one that gives them all a bad name. Many scientists find the idea one that is fun to think about, but they see it as not at all practical. This does not guarantee that there will never be an attempt to clone a man or that the day will never come when one is cloned. Much of the work in genetics makes the idea somewhat more realistic, but scientists predict that it will be at least ten to fifty years before a human might be cloned. Many scientists say that such a feat will never be accomplished or never be attempted. Many people from all walks of life feel that it never should be.

However, the idea of cloning a person fascinates so many people that stories about cloning people are popular. Why would anyone want to be cloned? Would it give satisfaction to someone on an ego trip? Would it be a way of feeling that one had another chance in life? Suppose Mr. Jones is cloned at the age of sixty-five. He believes that this will give him an heir who will have the same wonderful qualities that he thinks he has. But will the baby be just like Mr. Jones even if they have exactly the same genes?

The answer to that question is easy for almost anyone. In the first place, the baby will not have the same home environment. Where is his mother? Even the influence of natural body chemicals in the uterus of the surrogate mother, her diet, and other factors would influence the child to some degree. What has happened to make home life different in the period in which this child will grow to maturity from the time when Mr. Jones was a child? His school situation will be different. His friends will be different. Suppose the environment to which Mr. Jones' clone is exposed is such that the child becomes a juvenile delinquent. How would Mr. Jones feel about his heir if such were the case?

Someone has suggested that it would be necessary also to clone Mr. Jones' parents in order for his clone to be raised in the same way that Mr. Jones had been raised. When the clones of his parents were the same age as the parents were when Mr. Jones was born, a body cell, presumably preserved from Mr. Jones until the proper day, would be activated, according to this scenario. When the resulting baby was brought to full term in a surrogate mother's uterus, it could be delivered and placed in the home of the cloned parents. The parental environment would be somewhat the same, but not entirely. The whole idea of Mr. Jones' raising a child that is identical to himself soon becomes ridiculous.

Even if Mr. Jones could arrange to have himself cloned, would he be pleased to have a young image of himself, or would he feel that he had been supplanted and was no longer useful or important in this world? How would this clone feel about him? Could Mr. Jones provide the proper environment for good child care, or would he pressure the child to achieve, to accomplish more than he did?

Alvin Toffler in his famous book *Future Shock* suggests that making biological carbon copies of oneself is a fantastic possibility. However, he does not explain how it would "fill the world with twins" of those who are cloned. He suggests that it might help to resolve the old controversy about heredity and environment.

Certainly, the relationship between a clone and the person from whom he or she was cloned would be different from the usual parent-child relationship. How would the clone feel about the "parent?" In the story by Gene Wolfe "The Fifth Head of Cerberus," there are five generations of clones. Each one hates his "father" and murders him.

At the present time, no one else in all of the 4 billion people in the world is exactly like you. Even identical twins are raised in different environments. Family relationships, nutrition, and exercise may be different. But if cloning succeeded on a large scale, would the uniqueness of each individual cease to be important?

Figure 8-1. Although identical twins have the same genetic makeup, environmental influences can cause differences between them in personality.
Courtesy of Madeline Leggett

The idea of mass-produced clones fascinates people who dream of using them as slaves for unpleasant work. Actually, robots would be much cheaper and easier to produce. Suppose some country could produce a group of clones to fight their wars. They might do better with another kind of technology in which they did not have to plan a war twenty years in advance so that there would be time for the baby clones to grow up. Brainwashing or psychosurgery might serve them better. Such ideas belong to the kind of science fiction that can be expected to remain in the realm of wild fantasy.

Much the same is true about the stories of mass-produced Einsteins and Picassos. As in the case of Mr. Jones, the clones would grow up in environments different from those of the outstanding people they were expected to imitate. And what would be the psychological pressure on the children who were expected to accomplish so much? Even natural children of famous people often make special efforts not to follow the careers of their parents.

Many clone stories suggest they would have a special kind of rapport enabling one clone to know everything another is thinking. This type of relationship has been expressed in some fiction about twins, but in real life there has been no solid evidence to support such absolute telepathy between twins.

Identical twins have been the subject of many recent studies. Since they have two parents, they are disqualified as clones by some definitions. However, they have one parent in the technical sense: the single fertilized egg that gave rise to the two individuals with the same genetic makeup. According to some estimates there are 200 million people in the world today who are clones, and although they display many interesting similarities even when they have been separated from birth, they are not the kind of clones people fantasize about.

Although much has been written about the frightening aspects of cloning a man, the chances of such an event taking place appear slim. It seems probable that most of the people who purchased the book *In His Image: The Cloning of a Man* when it appeared in 1978 were either curious or eager to believe that the feat had been accomplished. Since the book was published as nonfiction by a reputable publisher, it was not

Figure 8-2. A clone of Einstein would not be exactly like the original because of different environmental influences. Cloning any human would raise serious ethical questions.
Institute for Advanced Study

surprising to find so many believers. The author, David Rorvik, claimed that a sixty-five-year-old industrialist had approached him for help in cloning an heir from cells of his own body. According to the author, the industrialist insisted that his identity be protected, so the whole procedure had to be carried out in secret. Although the book hints at the way the cloning was accomplished, it gives no specifics about the research procedures, claiming that this was necessary to protect those who were developing the cloning technique.

In His Image was "cloned" more than one hundred thousand times, but scientists objected strenuously to the book, calling it a hoax. This was especially true of the scientists whose work was mentioned in the footnotes. At least one scientist successfully sued the author and the publisher because his name was unfairly associated with the book.

Genetic engineering techniques are a powerful tool for manipulating nature, and even though this manipulation appears to hold the greatest hope for a cure for cancer, more food for a hungry world, and other benefits, it also challenges some deeply held feelings about the meaning of life and the value of family.

Scientists themselves were the first to voice concern about what was happening in experiments with recombinant DNA. Much of the controversy began at Cold Spring Harbor, New York, when Robert Pollack was teaching a summer course in 1971. At this time, a student described an experiment she was planning to do in Paul Berg's laboratory at Stanford University. After hearing this, Dr. Pollack became concerned about the possibility of new forms of life from such an experiment escaping from the laboratory and causing cancer in human beings. He telephoned Dr. Berg, expressing his concern. Paul Berg, an outstanding researcher in the field of genetics, felt that the risk was almost zero, but the prospect of creating an organism that might spread human cancer was so serious he decided not to carry out the proposed experiment. This was the beginning of a continuing debate over the responsibility and risks involved in genetic engineering experiments.

Figure 8-3. Work in genetic engineering is frequently done under a hood as a safety measure. *Stanford Medical Center News Bureau*

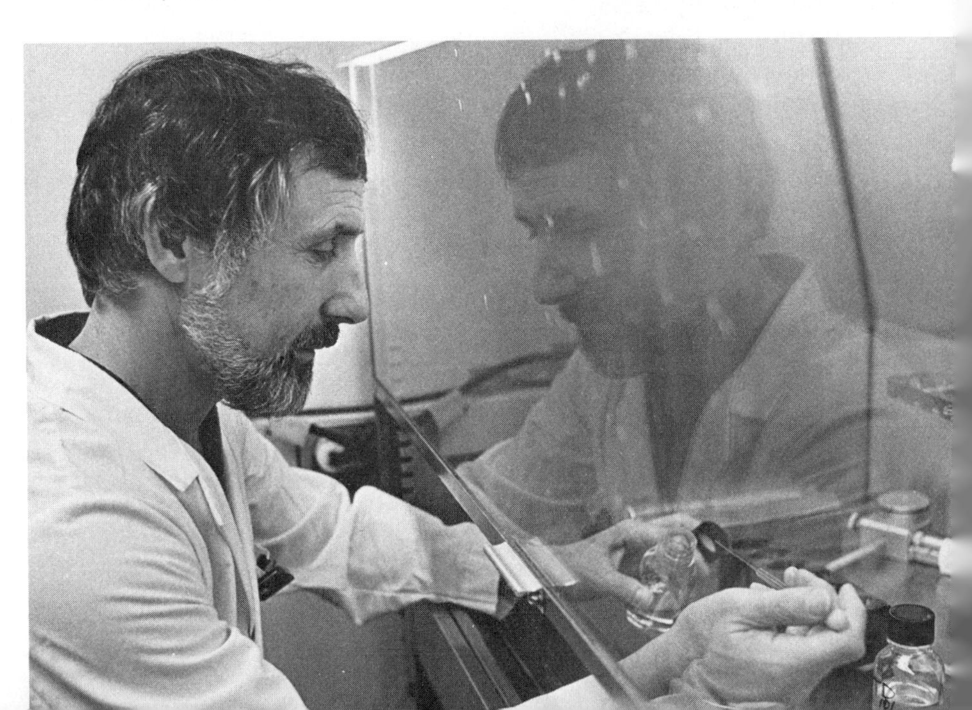

After this, small groups of scientists held meetings, and a letter was sent to leading scientific journals asking scientists to halt work on certain kinds of experiments. In February of 1975, a group of about 140 scientists from seventeen nations, some laymen, and a large number of journalists met at Asilomar Conference Center in Pacific Grove, California. Here they discussed the possible risks of certain kinds of work with recombinant DNA and what could be done about them. After some debate, it was decided that there was not enough information about the subject to be certain that there was absolutely no risk.

The National Institutes of Health that funded most of the DNA research set up a Recombinant DNA Advisory Committee to monitor research and draw up guidelines for experiments. These guidelines had to be followed to prevent any chance that a dangerous organism would escape and get into the environment.

Even before the conference at Pacific Grove, many scare scenarios appeared in the mass media. They ranged from microbes that erased the population of large areas of the world to Frankenstein monsters and beyond. Many people began to clamor for guidelines or even for a halt to work in the field of genetics.

Years later, after many people became informed that recombinant DNA occurred naturally in viruses, that nothing dangerous had appeared from the many laboratories, and that much was being done to benefit humanity, the fears subsided and the guidelines were relaxed. This does not mean that the work of the geneticists is being ignored.

When the U.S. Supreme Court handed down its decision in case number 79-136 in 1980, *Diamond* v. *Chakrabarty,* Dr. Ananda Chakrabarty became a part of legal history and a symbol for the storm of controversy over the perils and ethics of genetic engineering. Chakrabarty, Professor at the University of Illinois School of Basic Medical Sciences, predicts that controversy and speculation concerning genetic engineering have only just begun.

Chakrabarty's legal battle with the U.S. Patent Office began in 1972 when he and the General Electric Company, where he had been doing research, applied for a patent on a new

Figure 8-4. Dr. Ananda Chakrabarty and General Electric Co. pioneered efforts to patent a new life form developed through genetic engineering.
–University of Illinois Medical Center

form of life. The case concerned a variation of the common bacterium *Pseudomonas* which was created by genetic manipulation. The new form was developed for use in cleaning up oil spills or as a feed for cattle. At first, the patent was denied on the grounds that life forms cannot be patented. When the decision was overturned by the Supreme Court, many scientists hailed the news as a landmark decision that would spur research in genetic engineering by increasing an exchange of ideas.

This first patent on a new life form stirred renewed fears in the hearts of many people about a superrace of humans and about bacteria getting out of the lab and going wild. Dr. Chakrabarty feels certain that nothing like this would ever happen. He believes that much of their fear is fear of the unknown.

A close interaction between genetic engineers and public health workers may lead to new ways of treating water and sewage and to the creation of new strains of bacteria that can eat harmful chemicals. Dr. Chakrabarty has already developed bacteria that can totally decompose Agent Orange ingredient 2,4,5-T, turning the herbicide's toxic chemicals into carbon dioxide, water, and hydrochloric acid.

One of Dr. Chakrabarty's research projects is especially popular because it involves a major way of coping with the world food shortage. Human beings, unlike cattle and sheep, cannot eat grass and much other vegetation because their digestive systems cannot break down cellulose. Dr. Chakrabarty is trying to alter the intestinal bacterium *E. coli* through genetic engineering so that people could digest cellulose. Then millions of hungry people could add grass to their food supplies. No wonder many people are reconsidering their attitude toward genetic research.

In November 1982, a study team composed of members appointed by the National Council of Churches published the results of their two-year study of genetic engineering. After meeting with scientists, doctors, members of government agencies, consumer groups that monitor science and technology, and with representatives of pharmaceutical companies, they issued a report titled "Study Paper of Bioethical Concerns." Its authors suggested that the report be used as a basis for a policy to be drawn up over the next two years. It reminded us that we may not judge the recombinant DNA technology to be intrinsically good or evil, but we must be concerned how it is used.

The importance of genetic engineering is recognized as possibly being as crucial to our understanding of reality as Einstein's theory of the nature of matter. The National Council of Churches expressed concern as to whether there are broad risks to the environment in bringing new species into being, but recognized that "one valid goal of the life sciences and genetics in particular is the reduction of suffering. Surely the relief of suffering, in a world where suffering is both intense and unending, is an act of mercy."

In respect to the cloning of human beings, the report was especially definite: "We know of no existing laws or regulations to prohibit such experiments. We know of no goal legitimate enough to warrant such radical experimentation. There could be no margin for error. . . we ethically must call for perimeters to the use of genetic engineering."

The report ends on a positive note: "Nevertheless, it is exciting to gain new knowledge, to break new barriers. We celebrate the healing possibilities that are now before us through this new life technology. . .We are partners with one another in the ongoing journey for humankind; we are partners in the continuing creative process with God."

About the same time as the publication of the report by the National Council of Churches, the results of a two-year study by the President's Commission for the Study of Ethical Problems in Medicine and Biomedical and Behavioral Research were published. The commission accepted genetic engineering techniques as a powerful new tool if used responsibly. It recommended that a special group be formed to watch over areas of genetic engineering.

The month after the appearance of the two reports mentioned above, newspaper headlines announced the making of "mighty mouse." This announcement of the creation of a mouse not found in nature created much excitement when it was made in December 1982. Scientists were excited because of the possibility of using this technique to stimulate rapid growth in commercially valuable animals. They also realized that this was another step toward the use of this technology either to correct or to mimic certain genetic diseases that were being explored. The announcement also caused many people to become anxious about what was happening in the field of genetic engineering.

Scientists had been working for many years to take a trait from one species of mammal and transfer it into another species. Scientists at four American institutions cooperated to take a gene carrying the growth hormone from rats and link it with mouse genes so that the mice would accept it as their own. The mouse portion of the composite gene served as a switch to activate the rat gene.

Fused genes were injected into 170 fertilized mouse eggs and the eggs were placed in the reproductive systems of six foster mothers. After three weeks, the usual period of gestation,

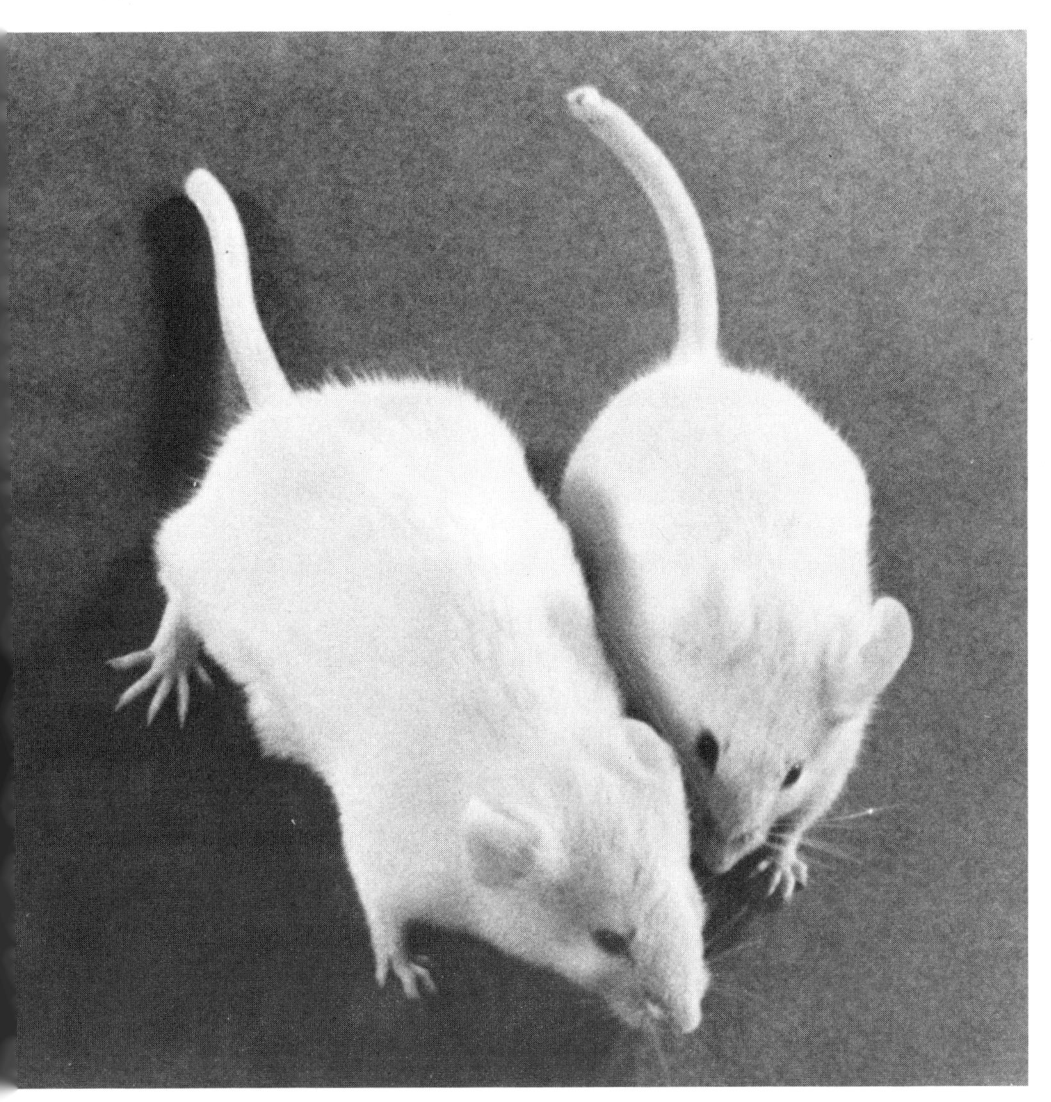

Figure 8-5. The mouse on the left contains a new gene made by fusing a mouse gene to a rat growth hormone gene. This mouse weighs 44 grams, while his sibling without the gene, on the right, weighs only 29 grams. In general, mice that express the gene grow two to three times as fast as controls and reach a size up to twice normal. *Photograph by R.L. Brinster*

twenty-one baby mice were born. It was soon apparent that some of the mice were growing two to three times faster than their brothers and sisters. Six of the seven mice that inherited the transplanted genes grew from 20 percent to 80 percent bigger than the others, and the gene lay dormant in one of the mice. The 170 transplant attempts produced only 21 mice, but this was a good record. In fact, the whole experiment was hailed as a dramatic feat of genetic engineering.

Why did these giant mice stir fear in the hearts of many people? No one was planning to breed supermice commercially. Even if one of the "mighty mice" escaped, an unlikely situation, no one worried that a whole new strain of larger mice would invade the fields of farmers and the homes of city dwellers. But questions were asked. If scientists can tamper with the genes of mice, what will they do with the genes of humans? Are they interfering with nature in a way that could be dangerous? Why do scientists tamper with nature?

The scientists who performed the experiments that produced the giant mice were concerned with learning more about the regulation of genes. What turns genes off and on? Such information, as mentioned earlier, is important in an understanding of why some cells go awry and produce cancer. This is one of the most important areas of cancer research.

Eye color has been changed in fruit flies by transplanting genes. And a rabbit gene for producing part of the hemoglobin found in blood was transplanted into mice, where a very small amount of it was produced in the mice. But transplanting genes from one animal to another is a very young art.

Decades may pass before it is possible to make use of gene therapy on a large scale, but faulty genetic messages are already being attacked. In May of 1982, it was reported that a team of scientists led by Dr. Yuet Wai Kan at the University of California had countered faulty genetic messages that caused a human blood disease known as beta thalassemia. The experiment involved snipping out a segment of the genetic material from cells of a patient who suffered from this disease, and replacing the segment with a gene that could tell cells to manufacture a substance that was needed. To prove that the technique would work, they put defective genes that cause

Figure 8-6. Dr. Yuet Wai Kan (right) is a pioneer in genetic work with thalassemia and sickle cell anemia. Here he and laboratory technologist Andree Dozy examine an autoradiograph of a sequence from human tRNA genes.
Robert Foothorap, UCSF

thalassemia and those with the man-made correction into frog egg cells. The treated cells produced the proper component of hemoglobin.

In December 1982, scientists reported that they had successfully altered the activity of genes in the human body in patients suffering from beta thalassemia. In essence, thalassemia patients cannot form healthy red blood cells. Blood disorders such as this and sickle cell anemia result from defects in the genes that control the production of hemoglobin, which carries oxygen to the cells. Scientists are experimenting with a drug that tends to activate certain otherwise defective genes. Researchers know that a chemical process called methylation causes some genes to turn off. They have temporarily overcome two inherited forms of anemia by using a drug to "switch on" a dormant gene.

Figure 8-7. The chance interchange of only 2 amino acids (arrow) out of 574 in this hemoglobin molecule changes its complex structure sufficiently to cause sickle cell disease. *Courtesy of Dr. Makio Murayama*

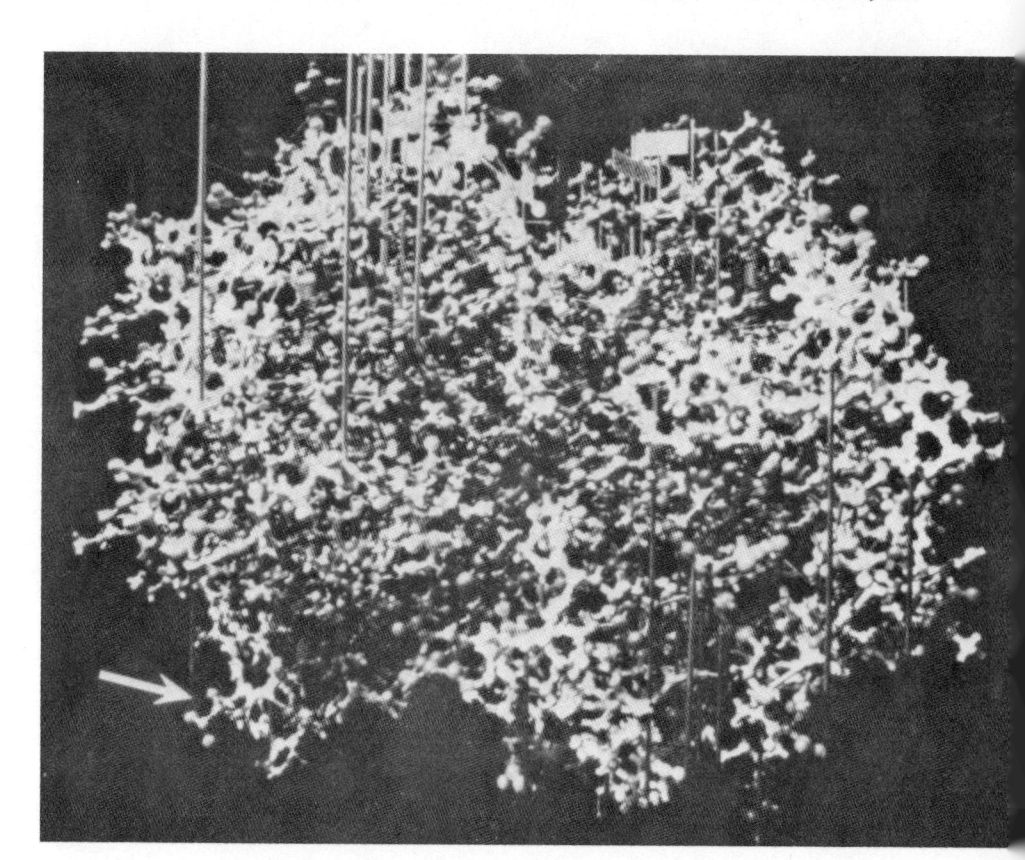

While researchers work toward manipulating genes to cure disease, many people fear tinkering with genetic material in humans. In 1983, new concern was voiced by a group of Roman Catholic, Jewish, and Protestant leaders. This group urged Congress to ban experiments that would change the characteristics passed on from one generation to the next. They made it clear that they were opposed to the manufacture of new forms of life, not the repair of physical defects in individuals. Many groups remind the public that any knowledge or invention can be used for good or evil. Fire and the wheel have long been used for both. If genetic research is suppressed in one country, it will be used in another.

Genetic therapy can be applied in two ways. One is through the control of genetic function; the other is the avoidance or replacement of defective or missing genes. Especially close scrutiny is being called for in the case of procedures that would create inheritable changes or changes that are aimed at enhancing normal people. Such possibilities may be a century away.

However, in the near future, scientists hope to replace defective genes in victims of two genetic diseases. In one case, the victims are deprived of an enzyme called PNP, which is necessary for the proper functioning of the human body. The lack of this enzyme destroys the immune system, bringing death at between the ages of two and nine from overwhelming infection. Many children with the defective gene succumb to chicken pox. Dr. David W. Martin, Jr. has located the healthy gene for making PNP in humans and has cloned it in the laboratory. He hopes to obtain permission to insert the healthy gene into the bone marrow of children who suffer from this disease so that new cells developed in the bone marrow will produce PNP.

Lesch-Nyhan patients suffer from a different gene-based enzyme deficiency. In this disease the enzyme is called HPRT, and its absence is responsible for many cruel symptoms, including mental retardation, crippling, kidney stones, and a compulsion for self-mutilation. There are children suffering from Lesch-Nyhan syndrome who bite themselves so severely that they would amputate their fingers if they were not restrained. Scientists are experimenting with a new technique using a virus to insert genes into mice bone marrow cells. Inder Verma and Dr. Theodore Friedmann inserted a normal human HPRT gene into a mouse leukemia virus. The virus, one of a class called retroviruses, had been modified so that it would not cause cancer. When mouse bone marrow cells infected with the gene-bearing virus were injected into the marrow of living mice, the marrow produced HPRT.

Scientists hope that one day they can use this gene therapy method to insert healthy HPRT genes into human bone marrow cells, which will then produce the crucial HPRT enzyme. Some families whose children suffer from this disease are willing to take the risk of such an experimental technique. Scientists are debating the wisdom of such procedures even though this kind of gene therapy would be limited to individuals and would not affect their offspring.

Everyone agrees that there is need to proceed with caution in any kind of gene therapy. Even those who accept the idea of taking a risk to cure cancer or help a suffering child are apt to raise their voices in protest when proposed gene therapy involves germ cells. Such changes, good or bad, would be inherited by future generations and would play a permanent role in the development of the human species.

Glossary

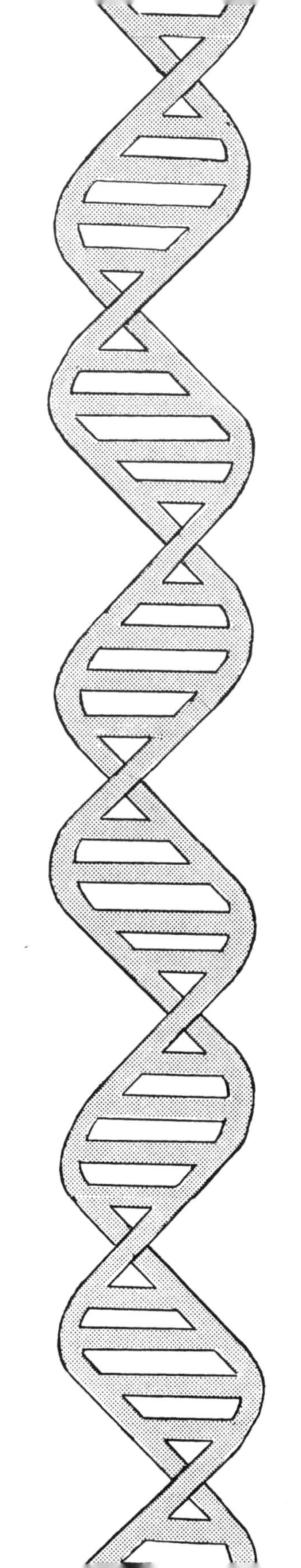

amino acids—the building blocks of proteins. There are twenty common amino acids that determine the character of proteins.

antibodies—proteins found in the blood which react with antigens (foreign substances) to make them harmless.

antigens—substances which when introduced into the body stimulate the production of antibodies. Each antibody will react with a specific antigen.

bacteriophage (phage)—a virus that multiplies in bacteria.

biotechnology—industrial processes that involve the use of biological systems.

callus—a cluster of plant cells that results from culturing tissue of a single plant.

cell fusion—joining two or more cells to form a single cell.

chromosome—a threadlike molecule of DNA wrapped around proteins. Chromosomes are found in the nucleus and contain most of the DNA in a cell. Genes are carried by the chromosomes.

clone—a group of identical cells or organisms that are descended from a common ancestor through asexual reproduction. All cells in a clone have the same genetic material and are copies of the original.

differentiation—biochemical and structural changes that groups of cells undergo to form a specialized tissue.

DNA (deoxyribonucleic acid)—the material in which the genetic instructions of all living things are embodied.

dominant gene—codes for a characteristic whose expression prevails over alternative characteristics for a given trait.

embryo transfer—implantation of an embryo from one animal or person into the oviduct or uterus of another, who is then the surrogate mother for the embryo.

enzyme—a protein that speeds or slows a chemical reaction. Enzymes control the rate of metabolic processes in an animal.

Escherichia coli (E. coli)—a bacterium whose natural habitat is the gut of humans and other warm-blooded animals. *E. coli* is a favorite subject for experimentation.

gene—the hereditary unit; instructions in the form of a stretch of DNA. An eye cell differs from a bone cell, for example, because a particular group of genes in a cell is active while the rest are somehow switched off.

gene therapy—switching off defective genes or inserting genes that specify the correct product.

gene mapping—determining the relative locations of different genes on a given chromosome.

hormones—"messenger" molecules that coordinate the actions of various tissues of the body; they produce a specific effect on the activity of cells remote from their point of origin.

hybrid—a new variety of plant or animal that results from combining two existing varieties. Hybrid DNA is made by joining pieces of DNA from different sources.

hybridoma—an immortal hybrid cell formed by the fusion of a cell that produces a single pure antibody with a cancer cell.

leukocytes—white cells of the blood.

lymphoma—a form of cancer that affects the lymph tissue.

messenger RNA—ribonucleic acid molecules that carry instructions from the DNA in the nucleus to the body of the cell where proteins are synthesized.

metabolism—the chemical processes involved in the maintenance of life and by which energy is made available.

mitochondria—structures in cells that serve as the "powerhouses" for the cells.

monoclonal antibody—a protein that recognizes only one kind of antigen; produced by cloning a hybridoma.

mutation—any change that alters the sequence of bases along the DNA, changing the genetic makeup, which causes loss or change of inherited traits.

myeloma—cancer of the blood-forming tissue; a primary tumor of the bone marrow.

nuclear transplant—an experimental technique in amphibians in which the nucleus from an embryo cell is transplanted into an egg cell whose nucleus has been removed.

parthenogenesis—reproduction in animals without the male fertilization of the eggs.

plasmids—small pieces of DNA that are carried by certain bacteria as extra minichromosomes.

proteins—products of gene expression; made up of amino acids, they are the functional and structural components of cells.

protoplast—a cell without a wall.

protoplast fusion—a means of changing genetic composition by joining two protoplasts or joining a protoplast with any of the components of another cell.

recessive gene—any gene whose expression is dependent on the absence of a dominant gene.

recombinant DNA—hybrid DNA produced by joining pieces of DNA from different sources.

recombinant DNA technique—a method of chemically cutting and splicing DNA.

restriction enzymes—enzymes that act like a scalpel to cut DNA molecules. Different restriction enzymes cut at different locations.

RNA (ribonucleic acid)—exists in three forms: messenger RNA, transfer RNA, and ribosomal RNA; assists in translating the genetic message of DNA into the finished protein.

tissue culture—a method of propagating healthy cells outside the organism.

viroid—a tiny snippet of RNA; the smallest known cause of infectious disease.

virus—an infectious agent that requires a host cell in order to reproduce itself. It is composed of either RNA or DNA wrapped in a protein coat.

Suggested Reading

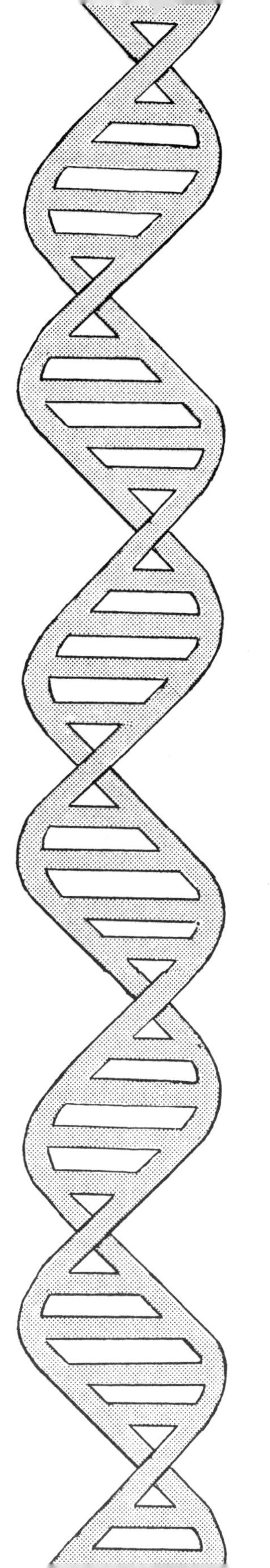

Anderson, J. Kerby, *Genetic Engineering: The Ethical Issues*, Grand Rapids, Michigan, Zondervan, 1982.

Cherfas, Jeremy, *Man-made Life: An Overview of the Science, Technology and Commerce of Genetic Engineering*, New York, Pantheon, 1983.

Cooke, Robert, *Improving on Nature: The Brave New World of Genetic Engineering*, New York, Times Books, 1978.

Farago, Peter and John Lagnado, *Life in Action: Biochemistry Explained*, New York, Alfred A. Knopf, 1972.

Glover, David M., *Genetic Engineering: Cloning DNA*, New York, Chapman and Hall, 1980.

Gonick, Larry and Mark Wheelis, *The Cartoon Guide to Genetics*, New York, Barnes and Noble/Harper and Row, 1983.

Harsanyi, Zsolt and Richard Hutton, *Genetic Prophecy: Beyond the Double Helix*, New York, Rawson-Wade, 1981.

Jackson, David and Stephen P. Stitch, *The Recombinant DNA Debate*, Englewood, New Jersey, Prentice-Hall, 1979.

Judson, Horace Freeland, *The Eighth Day of Creation*, New York, Simon and Schuster, 1979.

Lear, John, *Recombinant DNA: The Untold Story*, New York, Crown, 1978.

Luggre, David, *Life Manipulation: From Test Tube Babies to Aging*, New York, Walker and Co., 1979.

Restak, Richard, *Premeditated Man: Bioethics and the Control of Future Human Life*, New York, Viking Press, 1975.

Rosenfield, Israel, Edward Ziff and Borin van Loon, *DNA for Beginners*, Writers and Readers (distributed by W.W. Norton, New York), 1983.

Santas, M.A., *Genetics and Man's Future: Social and Moral Implications of Genetic Engineering*, Springfield, Illinois, C.C. Thomas, 1981.

Swertka, Eve and Albert Swertka, *Genetic Engineering*, New York, Watts, 1982.

Wade, Nicholas, *The Ultimate Experiment: Man-made Evolution*, New York, Walker and Co., 1977.

Wallace, Robert A., *The Genesis Factor*, New York, William Morrow, 1979.

Watson, James D. and John Tooze, *The DNA Story: A Documentary History of Gene Cloning*, San Francisco, W.H. Freeman, 1981.

Index

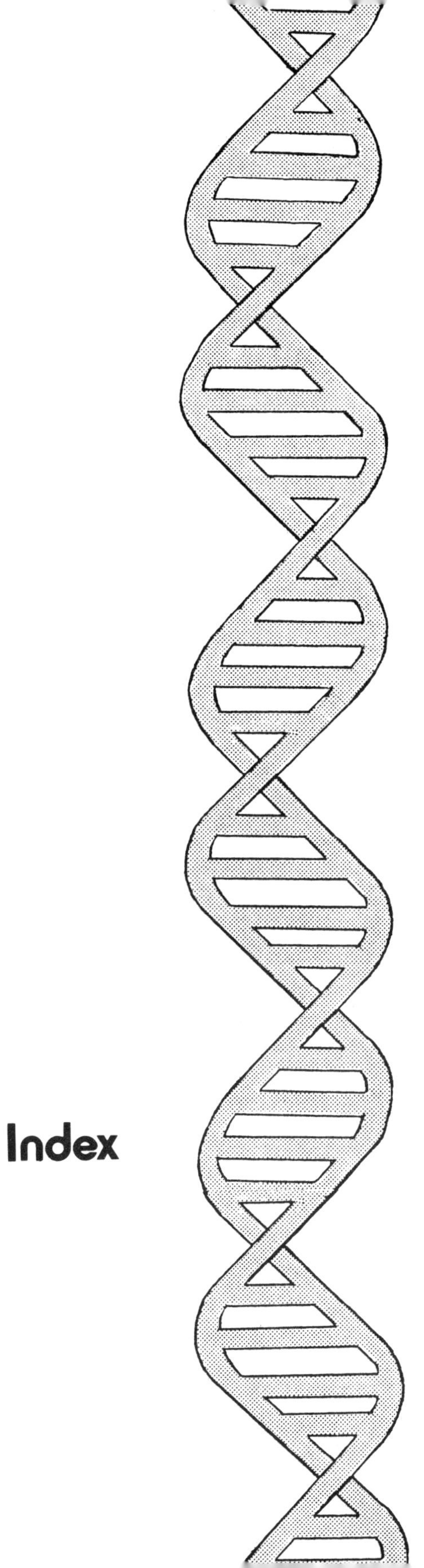